AI Application

人工智慧在流行趨勢研究的應用

in Fashion Trend

顏志晃 博士 (Chih-Huang Yen, Ph. D.) 著

英文版

Preface

　　在科技的突飛猛進下，每每都有令人驚豔的產品與流行的話題不斷，流行趨勢的信息一直圍繞在你我的生活中，每天早上一起床，準備換上班服裝，想著當個潮男或是扮文青，皆因流行信息左右你的思考；在出門之前的化妝打扮也無不思索著，現下最亮眼的脣色或是妝容，而無論是在裸肌或煙燻妝的選擇當下，顏色的魔力永遠是最神祕也最令人玩味的議題。

　　我一直被人戲稱都在賺女人的錢，職涯 20 年間，有 10 年做的是珠寶首飾設計，有 10 年做的是化妝品的研發設計，五年前因緣際會下進到成功大學工業設計研究所攻讀博士，我以色彩當成我的研究軸心，加上恩師的模糊理論，智能模式研究將其應用在人臉膚色的擷取與分類。這著作是我攻讀博士班的研究心血結晶，其間有業界中的流行趨勢探討，更是把色彩當成工程來做研究，自色立體的討論到人臉膚色的擷取，有著獨到的見解，特別是將人臉膚色擷取以六點做一個經濟快速的擷取模式，進而套用數學邏輯編寫 JAVA 程式，將上萬筆人臉資料輸入，經取得的資料再作分析，經由模糊聚類理論所編成的 APP，得到百萬筆數據資料，再純化成 18 個分類。

　　本書的完成要感謝成功大學 蕭世文 特聘教授，不是他的開導就沒有這著作的產生。半夜筆耕時，感謝家人的陪伴與包容支持，因為有這些溫暖的肩膀，本書方能順利完成。擁有健康和時間去投入有意義的事情，其實是幸福的。學而後知不足，本書在撰寫期間，其實尚有許多可再做延伸研究與發展的地方，期待各位讀者的迴響與不吝指教，也希望此書可以給讀者們對趨勢流行與色彩應用有另一層面的想法與見解。

AUG. 2020

SUMMARY

Consumer behavior is complicated. In the cosmetic market, personal intuition and fashion trends for colour selection are guidelines for consumers. A systematic method for female facial skin-color classification and an application in the makeup market are proposed in this study. In this paper, face recognition with a large number of images is first discussed. Then, an innovative method for colour capturing at selected points is presented and complexion-aggregated analysis is performed. This innovative method is an extension of face-recognition theory. Images in RGB format are converted to Lab-space format during data collection and then Fuzzy C-means theory is utilized to cluster and group the data. The results are classified and grouped in Lab value and RGB index. Two programs are created. The first program, "FaceRGB", captures colour automatically from images. The second program, "ColorFCM", clusters and groups the skin-color information. The results can be used to assist an expert system in the selection of customized colours during makeup and new-product development.

In the study case, with more than 10,000 Asian women photos, FaceRGB for automatic skin color capture, obtained skin color data and then divided by ColorFCM eighteen group results. In the end, the study combined with Merck's colour trend forecast, connected the clustering skin colour with six Merck's idea skin colour to do the pair, the results will be applied to cosmetic, and more clearly realize the value of research and the future development of the application.

Key words   facial skin colour; cosmetic; Fuzzy C-means.

ACKNOWLEDGEM`ENTS

First and foremost, I would like to show my deepest gratitude to my supervisor, Prof. Hsiao, a respectable, responsible and resourceful scholar, who has provided me with valuable guidance in every stage of the writing of this thesis. Without his enlightening instruction, impressive kindness and patience, I could not have completed my thesis. His keen and vigorous academic observation enlightens me not only in this thesis but also in my future study.

I shall extend my thanks to Shirley Lee for all her kindness and help. I would also like to thank all my teachers who have helped me to develop the fundamental and essential academic competence. My sincere appreciation also goes to George Chen (CEO from Tair Jiuh Group), who sponsored the tuition without return. I am proud of the stuff from Tair Jiuh. Also I will feedback to my company with the knowledge I learned.

Last but not least, I'd like to thank all my parents and wife, without their guidance and support I would not be here, especially my two lovely kids, for their encouragement and support.

TABLE OF CONTENTS

LIST OF SYMBOLS AND ABBREVIATIONS

FCM	Fuzzy C- Means
LO	Log-opponent chromaticity colour space
YC_bC_r	Y is Luma component , C_b is Blue-difference and C_r is Red-difference chroma components
CIE	Commission internationale de l'éclairage
Lab	L for lightness, a & b for the color opponents green–red and blue–yellow.
UCS	Uniform Color Space
HSV	Hue, Saturation and Value
CNN	Convolutional Neural Network
SCE	Skin Colour Extractor
FOS	Fiber of Spectrometer
EM	Expectation Maximization
P_D	Percentage of Distance
ΔE	CIELAB Delta E
$\Phi(x,y,z)$	Ellipsoid function

$N_{Taguchi}$	The minimum number of experiments from Taguchi method.
r	The correlation coefficient
F_{obj}	The objective function
$\omega_{i,j}$	The degree of a specific element
$\omega_K(x)$	A set of weight
C_K	The centroid of a cluster

CHAPTER 1

INTRODUCTION

For image processing, many researchers have been done on complexion applications, involving theories such as fuzzy theory, big data, pattern matching, and so on, in different fields. Applications for human face recognition have used neural network theory (F'eraud et al., 2001) and pattern recognition (Mario & Maltoni, 2000). Some face-recognition algorithms distinguish human faces from background in images using extracted features. In the feature-based approach (Brunelli & Poggio, 1993), algorithms analyze the relative positions, sizes, and/or shapes of the eyes, nose, cheekbones, and jaw. During image processing, algorithms for face-image normalization and feature compression are used to save vital data. All of these research areas are considered useful for complexion recognition. Many image applications for human faces contain feature-learning and matching functionalities. Currently, computer technology contributes faster calculating platforms on which to implement and practice these theories. One of the earliest successful systems is based on template-matching techniques (Brunelli, 2009) applied to a set of salient facial features, providing a sort of compressed face representation.

Many studies on skin colour focus on face recognition or try to determine the typology of people (Hsu et al., 2002). However, the cosmetic market may need to assist people in finding the right make-up colours for different conditions. However, determining a person's skin colour is also a large issue in cosmetic research. The main purpose of this study is to use females as an example to determine colour modes using an innovative method to extract features. A specific method is developed for clustering information based on Fuzzy c-means theory to obtain the grouping result.

1.1   Related research

Human face colour is vital information and is used in different applications, such as video surveillance, human-computer interfaces, face recognition, and face-image database management (Hsu et al., 2002).

Normally, studies of colour images with faces use colour information as the reference for human face detection, tracking, recognition, and so on. Therefore, choosing the "colour space" and determining how to build a "skin colour distribution model" to perform colour detection are the main problems. The author works for the cosmetic industry, and has spent several years' observing consumer behaviour. The reality is that it is hard for consumers to purchase appropriate colours due to insufficient knowledge, since not every end user is a make-up artist. The human face can be thought of as a canvas, and constructing a beautiful look is similar to painting like an artist. Thus, it is fundamental to determine one's face colour, which could be an indicator of a health condition. It also standard practice in cosmetics, since make-up is used to strengthen a person's appearance in some specific aspects.

In addition, objective evaluation of facial skin colour using a colorimeter is also popular and reliable (Caisey, 2006; Yoshikawa et al., 2009; Estanislao et al., 2003).

Table 1.1　A comparison for the major detection approaches published previously

Authors	Years	Approach	Features Used	Color Space	Findings
Tan et al.	2014	Automatic detection	Colour regions.	LO	Face detection
Yoshikawa et al.	2012	Evaluated visually	Facial image Skin colour	CIE L*a*b*	Colour captured

Zeng et al.	2011	Ellipsoid model	Colour region	CIE L*a*b*	Skin colour
Kakumanu et al.	2007	Skin model	Colour region	CIE L*a*b*	Face detection
Hsu et al.	2002	Detection algorithm	Face candidate	YC_bC_r	Face detection
Wu et al.	1999	Fuzzy Models	Colour regions	UCS	Face detection

Table 1.1 shows the approaches and findings of related studies. Obviously, skin-color capture is a new issue relative to other face-detection research. Yoshikawa et al. (2009) did some research on the relationship between white skin colour and health as a starting point to explore colour analysis. Colour spectrometers, along with expert interviews and cross matching, are used to finally characterize the complexion in terms of whiteness and health. These results illustrate that a higher-level process of face recognition affects whiteness perception, and the levels of facial-skin whiteness are classified using a facial skin-color distribution.

From Table 1.1, Tan (2014) used a cost concept to determine the face-detection affection. He proposed a new human-skin detection method that combines a smoothed 2-D histogram and a

Gaussian model; the method could automatically detect human skin in colour images. Qualitative and quantitative results on three standard public datasets and a comparison with state-of-the-art methods have shown the effectiveness and robustness of the proposed approach. We used a special colour space, called the log-opponent chromaticity colour space (Forsyth & Fleck, 1999). On the other hand, the trade-off between training-set size and classifier performance is an issue, so Jones et al. (2002) extracted skin colour with 2 billion pixels from 18,696 web images to optimize the performance.

Zeng & Luo (2011) provided three elliptical skin-color models for skin-color detection. The first one is a skin-color cluster using a single ellipse without lightness dependency; the second is a skin-color ellipse using different lightness levels to fit the shape of the skin-color cluster; and the third is an ellipsoid model to fit the skin-color boundary, which considerably simplifies the modeling and training processes. In terms of test results, the skin-color cluster with a single ellipse is easy in training and efficient in computation, while the formulation of lightness-dependent ellipses is more complex, and the computation of skin-color boundaries is less efficient; the ellipsoid skin-color model represents a compromise among the modeling complexity, computational efficiency, and detection accuracy.

Kakumanu et al. (2007) presented three approaches: (1) skin modeling and detection in various colour spaces; (2) a different skin-modeling method using a classification approach; and (3) various approaches that use skin-color constancy and dynamic-adaptation techniques. Kakumanu improved the skin-detection performance in dynamically changing illumination and environmental conditions. The research also indicated a factor with better skin-detection performance.

Hsu et al. (2002) created a face-detection algorithm based on skin-tone colour models and facial features in a digital-image system. The research used eye, mouth and boundary maps to identify possible facial candidates. This research overcomes the difficulty of detecting the low-luma and high-luma skin tones by using a nonlinear conversion to the YC_bC_r colour system.

Wu et al., (1999) provided two fuzzy models that could extract the skin- and hair-colour regions. These methods can increase the face-detection rate and reduce the mismatching rate. In addition, a successful example of face recognition under a complex background environment was presented.

In previous studies related to image perception, Hsiao et al.,

(1994; 1995; 1997) created a systematic method using fuzzy set theory for colour planning for product design. Additionally, some articles about the study of skin colour, such as a study on choosing clothes based on skin colour, appear in the literature. Hsiao et al., (2008) utilized skin-color detection with colour-harmony theory to create an expert system. The system could help people input their skin colour and generate better options for the colour of their clothing. The system includes an evaluation procedure that helps people indecision-making. Face recognition is a vital issue in existing studies, yet face colour is only used for identification during processing. Many research used face colour to filter out the range. Such colour capture-related technology research could also be extended to other applications based on science and technology in daily life.

Ikeda et al., (2014) used a gonio-spectrometer to explore the luminosity of skin coated with foundation. Artificial and human skin was used for the comparison. There is a large database indicating the relationship between the colour space and skin colour. The luminosity is evaluated for cheeks and foreheads coated with powder foundation under different observation conditions using another experimental device. Finally, all luminosity scores are analyzed in colour spaces to identify the colour region related to luminosity perception.

1.2   Market Assessment from Merck

Merck (2016) is comprised of three business sectors: Healthcare (Biopharma, Consumer Health, Allergopharma, and Biosimilars), Life Science and Performance Materials.

Merck's product range includes a broad selection of special-effect pigments and functional fillers that enable you to incorporate every new trend and colour scheme into products. The cosmetic industry benefits from Merck's cosmetic pigments, which can be used in all types of cosmetics and personal-care applications to add colour, luster, shimmer and shine. The functional fillers improve the feel of the skin, along with the wear properties and the application features of the finished product. The Yearly Fashion-Market Assessment from Merck is like a bible for the fashion and beauty industry.   This paper will present a result from the study and present the connection between skin colour and cosmetic trends.

1.3   Outline of this study

Capturing the female facial-skin colour and performing the

clustering are the main purposes of this study. Toward these goals, two JAVA programs, "FaceRGB" and "ColorFCM", are prepared and also contain a colour-space conversion. Figure 1.1 illustrates the overall concept of all the procedures in the study.

"FaceRGB" is JAVA program that contains an innovative way to perform the colour detection from 6 points for each face in big data. As for the Colour System, the main function is to capture the sense of colour RGB data. The formula to convert the colour spectrum to CIELAB Lab is implemented in the program as well. Table 1.1 shows the approach of Yoshikawa et al., (2007), Zeng & Luo (2011) and Kakumanu et al., (2007) with CIELAB. The use of logarithms to reduce illumination change to simple translations of coordinates is more representative and reliable than other approaches; therefore, this paper determines colour space with CIELAB.

The second program, "ColorFCM", is prepared for colour clustering based on fuzzy theory. It classifies and clusters numerous data. The research process is shown in Figure 1.1, and it is described as follows.

First, collect as much data as possible through the internet or from digital devices (such as scanners, cameras or smart phones).

Input the picture into program "FaceRGB" for human-face

extraction via face recognition; on the other hand, facial features are generated for information analysis.

Perform the identification of six points based on their relative positions.

After colour capture of the six points for the action, a debugging process in the program is used to delete outliers.

The RGB information is captured by the inner program and automatically converted into Lab values.

The ColorFCM program is adopted for clustering the colour data (Lab) gathered using the FaceRGB program. This paper uses fuzzy theory to do the clustering.

Each clustering group has designated number, or an iterative clustering number is automatically generated by the program.

Grouping results can be presented in the 3D-space scatter plot and the seat instructions are marked.

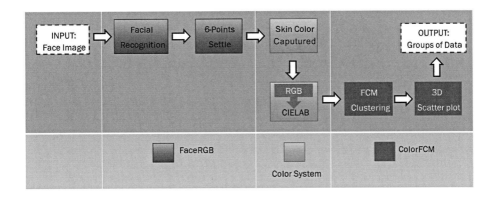

Figure 1.1　The research procedure

CHAPTER 2
LITERATURE REVIEW

2.1　The choice for colour space

This is well established that the distribution of colours in an image is often a functional element. An image could be represented using a number of different colour-space models (e.g., RGB, YC_bC_r (Ikeda et al., 2014)), and HSV (Vadakkepat et al.,2008), and it is necessary to choose the appropriate colour space for modeling facial skin colour.

A colour space may be arbitrary, with particular colours assigned to a set of physical colour swatches along with corresponding names or numbers, such as with the Pantone system, or structured mathematically, as is the RGB colour model. A colour model is an abstract mathematical model that assists the digital system in describing and illustrating the normal analog picture (e.g., triples in RGB or quadruples in CMYK). According to the artificial definition and design, a digital image is constructed with numerous pixels and these pixels each contain two pieces of information. The first piece of information is the colour, which is defined by a specific colour model; the second piece of information is the associated mapping-position information. The CIELAB or CIEXYZ colour spaces are frequently used to define the colour space because these two methods are specifically designed to illustrate the colours and spectra that most human beings can see.

2.1.1 RGB colour model

The RGB model is constructed with red light, green light, and blue light; mixtures of these three light beams, and their light spectra, are added wavelength for wavelength to create the final colour spectrum (Boughen, 2003).

Colours are light reflections from objects and this type of reflective light has different illuminations and spectra. The RGB model is one illustration approach, and there are also other colour models such as CIELAB and Y'UV. A three-dimensional volume is described by treating the component values as ordinary Cartesian coordinates in a Euclidean space. For the RGB model, Figure 2.1(a) represents a cube using non-negative values within the 0–1 range, assigning black to the origin at the vertex (0, 0, 0). Increasing intensity values run along the three axes, up to white at the vertex (1, 1, 1), and diagonally to the opposite colour, black. The RGB cube is rendered in Figure 2.1(b).

An RGB triplet (r,g,b) represents the three-dimensional coordinates of the point of the given colour within the cube, its faces or along its edges. This approach allows the computation of the colour similarity of two given RGB colours by simply calculating the distance between them: the shorter the distance, the higher the similarity. Out-of-gamut computations can also be performed in this way.

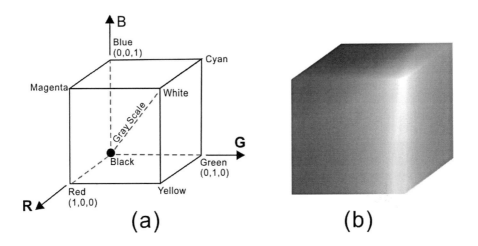

Figure 2.1　RGB model

2.1.2 CIELAB colour model

The colour space was originally defined by CAE and specified by the International Commission on Illumination. This colour model is device independent, and it provides the opportunity to communicate different colours across different devices. A Lab colour space is a colour-opponent space with dimension **L** for lightness and **a** and **b** for the colour-opponent dimensions, based on nonlinearly compressed (e.g., CIE XYZ colour-space) coordinates. The terminology originates from the three dimensions of the Hunter 1948 colour space (Hunter, 1985), which are L, a, and b. However, *Lab* is now more often used as an informal abbreviation for the L-a-b representation of the CIE 1976 colour space (or CIELAB, described below).

The difference between the original Hunter and CIE colour coordinates is that the CIE coordinates are a cubic-root transformation of the colour data, while the Hunter coordinates are a square-root transformation (Hunter, 1985). Other examples of colour spaces with Lab representations include the CIE 1994 colour space and the CIE 2000 colour space.

2.1.3 RGB & CIELAB conversions

Since RGB colour models are device dependent, there is no

simple formula for conversion between RGB values and *L*a*b**. The RGB values must be transformed via a specific absolute colour space. This adjustment will be device dependent, but the values resulting from the transform will be device independent. After a device-dependent RGB color space is characterized, it becomes device-independent.

In the calculation of sRGB from CIE XYZ is a linear transformation, which may be done by a matrix multiplication. Refer to Eqs. (1)-(2), it presents that these linear RGB values are not the final result as they have not been adjusted for the gamma correction. sRGB was designed to reflect a typical real-world monitor with a gamma of 2.2, and the following formula transforms the linear RGB values into sRGB. Let C_{linear} be R_{linear}, G_{linear}, or B_{linear}, and C_{srgb} be R_{srgb}, G_{srgb} or B_{srgb}. The sRGB component values R_{srgb}, G_{srgb}, B_{srgb} are in the range 0 to 1. (A range of 0 to 255 can simply be divided by 255.0).

$$C_{linear} = \begin{cases} \dfrac{C_{srgb}}{12.92}, & C_{srgb} \leq 0.04045 \\ \left(\dfrac{C_{srgb}+a}{1+a}\right)^{2.4}, & C_{srgb} > 0.04045 \end{cases}$$

$$(1)$$

where $a = 0.055$ and where *C* is *R*, *G*, or *B*.

Followed by a matrix multiplication of the linear values to get XYZ:

$$\begin{bmatrix} X \\ Y \\ Z \end{bmatrix} = \begin{bmatrix} 0.4124 & 0.3576 & 0.1805 \\ 0.2126 & 0.7152 & 0.0722 \\ 0.0193 & 0.1192 & 0.9505 \end{bmatrix} \begin{bmatrix} R_{linear} \\ G_{linear} \\ B_{linear} \end{bmatrix}$$

(2)

These gamma-corrected values are in the range 0 to 1. If values in the range 0 to 255 are required. The values are usually clipped to the 0 to 1 range. This clipping can be done before or after this gamma calculation (Green, 2002).

2.2   Taguchi Methods

Taguchi method is used to make designed product has stable quality, small fluctuation, and makes production process insensitive to every kind of noise. In the product design process, it uses relations of quality, cost and profit to develop high-quality product under condition of low cost. Taguchi method thinks the profit of product development can use internal profit of enterprise and social loss to measure, enterprise internal profit indicates low cost under condition with the same functions, social profit uses effect on human after product entering consumption field as the measurement index. This research uses Taguchi method, its main aim is to find out the optimization of skin color point, because point distribution has many probabilities, it can find out the optimal point model through calculation of Taguchi method.

If Taguchi's designs aimed to allow greater understanding of variation than did many of the traditional designs from the analysis of variance. Taguchi contended that conventional sampling is inadequate here as there is no way of obtaining a random sample of future conditions. In Fisher's design of experiments and analysis of variance, experiments aim to reduce the influence of nuisance factors to allow comparisons of the mean treatment-effects (Hardin et al., 1993). Variation becomes even more central in Taguchi's thinking.

Taguchi proposed extending each experiment with an "outer array" (possibly an orthogonal array); the "outer array" should simulate the random environment in which the product would function. This is an example of judgmental sampling. Many quality specialists have been using "outer arrays".

Later innovations in outer arrays resulted in "compounded noise." This involves combining a few noise factors to create two levels in the outer array: First, noise factors that drive output lower, and second, noise factors that drive output higher. "Compounded noise" simulates the extremes of noise variation but uses fewer experimental runs than would previous Taguchi designs.

Many of the orthogonal arrays that Taguchi has advocated are saturated arrays, allowing no scope for estimation of interactions. This is a continuing topic of controversy. However, this is only true for "control factors" or factors in the "inner array". By combining an inner array of control factors with an outer array of "noise factors", Taguchi's approach provides "full information" on control-by-noise interactions, it is claimed. Taguchi argues that such interactions have the greatest importance in achieving a design that is robust to noise factor variation. The Taguchi approach provides more complete interaction information than typical fractional factorial designs, its adherents claim. Followers of Taguchi argue that the designs offer rapid results and that interactions can be eliminated by proper choice of quality characteristics. That notwithstanding, a "confirmation experiment" offers protection against any residual interactions. If the quality characteristic represents the energy transformation of the system, then the "likelihood" of control factor-by-control factor interactions is greatly reduced, since "energy" is "additive"(Pukelsheim, Friedrich, 2006).

2.2.1 Basic concept

Taguchi has envisaged a new method of conducting the

design of experiments which are based on well-defined guidelines. This method uses a special set of arrays called orthogonal arrays. These standard arrays stipulates the way of conducting the minimal number of experiments which could give the full information of all the factors that affect the performance parameter. The crux of the orthogonal arrays method lies in choosing the level combinations of the input design variables for each experiment.

2.2.2　Properties of an orthogonal array

The orthogonal arrays has the following special properties that reduces the number of experiments to be conducted.

1. The vertical column under each independent variables of the above table has a special combination of level settings. All the level settings appears an equal number of times. For L9 array under variable 4 , level 1 , level 2 and level 3 appears thrice. This is called the balancing property of orthogonal arrays.

2. All the level values of independent variables are used for conducting the experiments.

3. The sequence of level values for conducting the experiments shall not be changed. This means one cannot conduct experiment 1 with variable 1, level 2 setup and experiment 4

with variable 1 , level 1 setup. The reason for this is that the array of each factor columns are mutually orthogonal to any other column of level values. The inner product of vectors corresponding to weights is zero. If the above 3 levels are normalized between -1 and 1, then the weighing factors for level 1, level 2 , level 3 are -1 , 0 , 1 respectively. Hence the inner product of weighing factors of independent variable 1 and independent variable 3 would be

$$(-1 * -1+-1*0+-1*1)+(0*0+0*1+0*-1)+(1*0+1*1+1*-1)=0$$

The design of experiments using the orthogonal array is, in most cases, efficient when compared to many other statistical designs. Referring the Eq. (3), the minimum number of experiments that are required to conduct the Taguchi method can be calculated based on the degrees of freedom approach.

$$N_{Taguchi} = 1 + \sum_{i=1}^{NV}(L_i - 1) \tag{3}$$

For example, in case of 8 independent variables study having 1 independent variable with 2 levels and remaining 7 independent variables with 3 levels (L_{18} orthogonal array), the minimum number of experiments required based on the above equation is 16. Because of the balancing property of the orthogonal arrays, the total number of experiments shall be multiple of 2 and 3. Hence the number of experiments for the above case is 18.

2.2.3　Assumptions of the Taguchi method

The additive assumption implies that the individual or main effects of the independent variables on performance parameter are separable. Under this assumption, the effect of each factor can be linear, quadratic or of higher order, but the model assumes that there exists no cross product effects (interactions) among the individual factors. That means the effect of independent variable 1 on performance parameter does not depend on the different level settings of any other independent variables and vice versa. If at time, this assumption is violated, then the additivity of the main effects does not hold, and the variables interact.

2.2.4　Designing an experiment

The design of an experiment involves the following steps

1.　Selection of independent variables

2.　Selection of number of level settings for each independent variable

3.　Selection of orthogonal array

4.　Assigning the independent variables to each column

5.　　Conducting the experiments

6.　　Analyzing the data

7.　　Inference

2.3　Fuzzy C-means

Data clustering is the process of dividing data elements into classes or clusters so that items in the same class are similar and items in different classes are dissimilar. Depending on the nature of the data and the purpose for which the clustering results are going to be used, different measures of similarity may be used to place items into classes, where the similarity measure controls how the clusters are grouped. Some examples of measures that can be used in clustering include distance, connectivity, and intensity. In hard clustering, data are divided into distinct clusters, where each data element belongs to exactly one cluster. In fuzzy clustering (also referred to as soft clustering), data elements belong to more than one cluster, and a set of membership levels is associated with each element.

The membership levels indicate the strength of the association between a data element and a particular cluster. Fuzzy clustering is the process of assigning these membership levels, and then using them to identify data elements to one or more clusters (Bezdek, 2013). One of the most widely used fuzzy clustering algorithms is the Fuzzy C-Means (FCM) Algorithm.

The FCM algorithm attempts to partition a finite collection of elements, $X=\{x_1, x_2, ..., x_n\}$, into a collection of c fuzzy clusters with respect to some given criterion. Given a finite set of data, the algorithm returns a list of C cluster centres, $C = \{c_1, c_2, ..., c_n\}$, and a partition matrix, $W = \omega_{i,j} \in [0,1]$, $i = 1, ..., n$, $j = 1, ..., c$, where each element $\omega_{i,j}$ is the degree of a specific element x_i belonging to corresponding cluster c_j. Referring to Eqs.(4)-(5), the FCM aims to minimize an objective function (Bezdek, 2013):

$$F_{obj} = \underset{C}{\text{argmin}} \sum_{i=1}^{n} \sum_{j=1}^{c} \omega_{ij}^{m} \left\| x_i - c_j \right\|^2,$$

$$(4)$$

$$\text{where} \quad \omega_{ij} = \frac{1}{\sum_{k=1}^{c} \left(\frac{\left\| x_i - c_j \right\|}{\left\| x_i - c_k \right\|} \right)^{\frac{2}{m-1}}}$$

$$(5)$$

In fuzzy clustering, a point has a different degree of belonging to different clusters. In scoring, the point's close the cluster centre obtain higher grades, and those far away from the centre obtain lower grades (Nock & Nielsen, 2006). An overview and comparison of different fuzzy clustering algorithms is available.

Any point x has a set of weights $\omega_K(x)$ for the k-th cluster; see Eq. (6). In fuzzy c-means, the centroid of a cluster is the mean value calculated from all points, weighted by their degree of belonging to the cluster:

$$C_K = \frac{\sum_X \omega_K(x)^m x}{\sum_X \omega_K(x)^m}. \tag{6}$$

The weight $\omega_K(x)$, relates to the opposite direction from the cluster centre, as calculated ion the previous pass. It also depends on parameter m which controls how much weight is given to the closest center (Nock & Nielsen, 2006).

The procedure of the fuzzy c-means algorithm is as follows.

1. Choose a number of clusters.
2. Randomly assign weights to each point to describe its degrees of membership in the clusters.
3. Repeat the steps until the algorithm has converged, that is, until the changes in the coefficients between two iterations are no more than ε, the given sensitivity threshold.
4. Compute the centroid for each cluster using Eq. (6).

5. For each point, compute its new weights in the clusters using Eq. (6) again.

2.4　Facial recognition system

In the study, the author capture the face colour base on the result of facial recognition. A face recognition system is a computer application capable of identifying or verifying a person from a digital image or a video frame from a video source. One of the ways to do this is by comparing selected facial features from the image and a face database. It is typically used in security systems and can be compared to other biometrics such as fingerprint or eye iris recognition systems. Recently, it has also become popular as a commercial identification and marketing tool (Brunelli, 2009).

Some face recognition algorithms identify facial features by extracting landmarks, or features, from an image of the subject's face. For example, an algorithm may analyze the relative position, size, and/or shape of the eyes, nose, cheekbones, and jaw. These features are then used to search for other images with matching features. Other algorithms normalize a gallery of face images and then compress the face data, only saving the data in the image that is useful for face recognition (Brunelli & Poggio, 1993). A probe image is then compared with the face data. One of the earliest successful systems is based on template matching techniques applied to a set of salient facial features, providing a sort of compressed face representation.

Recognition algorithms can be divided into two main approaches, geometric, which looks at distinguishing features, or photometric, which is a statistical approach that distills an image into values and compares the values with templates to eliminate variances.

Popular recognition algorithms include principal component analysis using Eigenfaces, linear discriminant analysis, elastic bunch graph matching using the Fisherface algorithm, the hidden Markov model, the Multilinear subspace learning using tensor representation, and the neuronal motivated dynamic link matching.

A newly emerging trend, claimed to achieve improved accuracy, is three-dimensional face recognition. This technique uses 3D sensors to capture information about the shape of a face.

One advantage of 3D face recognition is that it is not affected by changes in lighting like other techniques. It can also identify a face from a range of viewing angles, including a profile view. Three-dimensional data points from a face vastly improve the precision of face recognition. 3D research is enhanced by the development of sophisticated sensors that do a better job of capturing 3D face imagery. The sensors work by projecting structured light onto the face. Up to a dozen or more of these image sensors can be placed on the same CMOS chip—each sensor captures a different part of the spectrum.

Even a perfect 3D matching technique could be sensitive to expressions. For that goal a group at the Teknion applied tools from metric geometry to treat expressions as Isometrics. A company called Vision Access created a firm solution for 3D face recognition. A new method is to introduce a way to capture a 3D picture by using three tracking cameras that point at different angles; one camera will be pointing at the front of the subject, second one to the side, and third one at an angle. All these cameras will work together so it can track a subject's face in real time and be able to face detect and recognize (Socolinsky et al., 2004)

Face^{++} is the new visual service platform subordinated to Beijing Megvii Co., Ltd, which provides simple, easy, platform compatible and new visual service with strong functions.

With the birth of Kinect and GOOLE Glass, machine vision and human-computer interaction technology will become to be the core driving force of the next IT revolution wave. Multui-media information and vision technology have quietly changed people's life, for example it is used in social media Facebook etc., which is indicated by Figure 2.2. Human face has the richest information, it can quickly connect relevant visual information as for users, and its huge values are beyond all doubt.

Face[++] team is expert in researching and developing the best facial detection, identification, and analysis and reconstruction technology in the world, which integrates machine vision, machine learning, big data collection and 3D graphic learning technology, Sun et al.,(2013) put forward using Pyramid CNN (Convolutional Neural Network) as base and there are many CNN on every layer, it is respectively corresponds to area point of every input image, it is completed by all the area point characteristic assemble in all the layers, while choice of area point just relies on detection and correction of human facial characteristic.

Figure 2.2　Face recognition example in FaceBook

CHAPTER 3
IMPLEMENTATION METHOD

From the web, more than ten thousand female photos were collected to detect facial skin colour, by obtaining the skin-color data, clustering and determining the colour types from the images. From beginning, the study will make some pre-test to verify the hypothesis. Figure 3.1 show the procedure for the designing an experiment.

The design of an experiment (pre- test) involves the following steps

1. Create Skin Colour Extractor (SCE) Program based on Ellipsoid Skin-Colour model.
2. Get the Skin-Colour Data
3. Make the optimization with Taguchi method
4. Hypothesis for 6-points capture.
5. Verification for 6-points to detect facial colour.
6. Spectrometer Application
7. Test design

The details of the above steps are given below.

3.1   6 points colour detection

The objective of this program is to extract skin colour from photographic images. It is assumed that the typical image-processing software (Photoshop, CorelDraw, etc.) is employed, as shown in the different steps in Figure 3.1. A step-by-step procedure, shown in Figure 3.2(a), is used to read each file. Figure 3.2(b) shows that the background has been filtered out to completely reveal the face shape. It is possible for the user to capture the skin colour manually, and most of the position is decided based on intuition. Figure 3.2(c) presents the 6 points on the face, and the results of the colour detection are indicated with numbers in Figure 3.2(d). Then, the average values of the 6 data points can be calculated as shown in Figure 3.2(e). The flow chart illustrates the process procedure for the FaceRGB program.

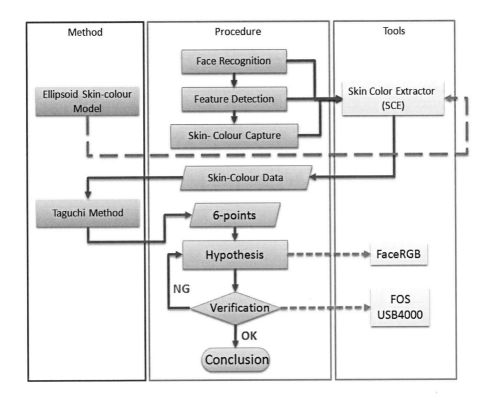

Figure 3.1　The procedure of the pre-test to verify the
6-points capture hypothesis.

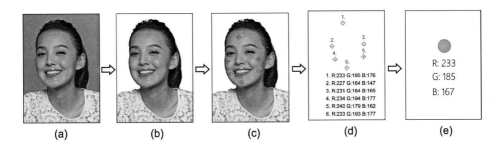

Figure 3.2　The traditional method to detect the skin
colour from image.

3.1.1   Hypothesis _ Captured colour from 6 points

In the research of Yoshikawa et al., (2007) for the skin-whitening project for Japanese women in SHISEIDO Co., Ltd., Tokyo, Japan, a spectrometer is used for capturing skin colour. Its measurement is as the forehead, cheek and under cheek. This study extends this approach to the six points for the basic operations of facial skin-color capture. In the study, Captured colour from 6 points is a hypothesis base on the lecture. And author will make some experiment to verify.

3.1.1.1   Skin Color Extractor,   SCE

Zeng et al., (2011) had the studies in human skin color luminance dependence cluster shape discussed in the Lab color space. The cluster of skin colors may be approximated using an elliptical shape. Let $X_1, \ldots, X_n$ be distinctive colors (a vector with two or three coordinates) of a skin color training data set and $f(X_i)=f_i (i=1, \ldots, n)$ be the occurrence counts of a color, X_i. An elliptical boundary model $\Phi(X) = (X, \Psi, \Lambda)$ is defined as

$$\Phi(X) = [X - \Psi]^T \Lambda^{-1} [X - \Psi], \tag{7}$$

where Ψ and Λ are given by

$$\Psi = \frac{1}{n}\sum_{i=1}^n X_i \;, \tag{8}$$

$$\Lambda = \frac{1}{n}\sum_{i=1}^n f_i (X_i - \mu)(X_i - \mu)^T \;, \tag{9}$$

where $N = \frac{1}{n}\sum_{i=1}^{n} f_i$ is the total number of occurrences in a training data set and $\mu = \frac{1}{n}\sum_{i=1}^{n} f_i X_i$ is the mean of color vectors.

To consider the lightness dependency of the shape of skin cluster, the cluster of skin colors in a lightness-chrominance color space may be modeled with an ellipsoid. In a three dimensional (3D) color space, X is expressed as

$$X = \begin{pmatrix} x \\ y \\ z \end{pmatrix},$$

and Λ^{-1} is represented in a matrix form,

$$\Lambda^{-1} = \begin{pmatrix} \lambda_{00} & \lambda_{01} & \lambda_{02} \\ \lambda_{10} & \lambda_{11} & \lambda_{12} \\ \lambda_{20} & \lambda_{21} & \lambda_{22} \end{pmatrix} \tag{10}$$

$\Phi(x)$ in Eq. (7) can be reorganized as

$$\Phi(x, y, z) = \lambda_{00}(x - x_0)^2 + (\lambda_{01} + \lambda_{10})(x - x_0)(y - y_0) + $$
$$(\lambda_{02} + \lambda_{20})(x - x_0)(z - z_0) + \lambda_{11}(y - y_0)^2 + (\lambda_{12} + \lambda_{21})(y - $$
$$y_0)(z - z_0) + \lambda_{22}(z - z_0)^2 \tag{11}$$

According to Eq. (9),

$$\Lambda = $$
$$\frac{1}{n}\sum_{i=1}^{n} f(x_i, y_i, z_i) \times$$
$$\begin{pmatrix} (x_i - x_0)^2 & (x_i - x_0)(y_i - y_0) & (x_i - x_0)(z_i - z_0) \\ (x_i - x_0)(y_i - y_0) & (y_i - y_0)^2 & (y_i - y_0)(z_i - z_0) \\ (x_i - x_0)(z_i - z_0) & (y_i - y_0)(z_i - z_0) & (z_i - z_0)^2 \end{pmatrix} \tag{12}$$

Comparing Eq. (11) to Eq. (12), $\lambda_{01} = \lambda_{10}$ and $\lambda_{21} = \lambda_{12}$
The ellipsoid function (11) can be written as:

$$\Phi(x, y, z) = u_0(x - x_0)^2 + u_1(x - x_0)(y - y_0) + u_2(y - y_0)^2 +$$
$$u_3(x - x_0)(z - z_0) + u_4(y - y_0)(z - z_0) + u_5(z - z_0)^2 \ ,$$

$$(13)$$

where $u_0 = \lambda_{00}$, $u_1 = \lambda_{01} + \lambda_{10}$, $u_2 = \lambda_{11}$, $u_3 = \lambda_{02} + \lambda_{20}$,
$u_4 = \lambda_{12} + \lambda_{21}$, and $u_5 = \lambda_{22}$.

$$\Phi = \lambda_{00}(r - r_0)^2 + (\lambda_{01} + \lambda_{10}) \ (r - r_0)(g - g_0) + (\lambda_{02} + \lambda_{20})(r -$$
$$r_0)(b - b_0) + \lambda_{11}(g - g_0)^2 + (\lambda_{12} + \lambda_{21})(g - g_0)(b - b_0) +$$
$$\lambda_{22}(b - b_0)^2 \ (14)$$

$$\Phi(r, g, b) = u_0(r - r_0)^2 + u_1(r - r_0)(g - g_0) + u_2(g - g_0)^2 +$$
$$u_3(r - r_0)(b - b_0) + u_4(g - g_0)(b - b_0) + u_5(b - b_0)^2 = \rho \ (15)$$

On the basis of human face identification, it uses characteristic point to make setting of relative position, applies skin color ellipse model, CNN of human face identification, and uses Java program to compile skin color extractor, its short form is SCE.

3.1.1.2 The instruction operation for Skin Color Extractor (SCE)

Figure 3.3 is the operation instruction of SCE, click (1) to open file, insert file, will see(2) figure display area indicates this file image and (3) file name, click (4) Detect Feature option on button area, image will automatically generate blue line, which indicates it has detected color point of right cheek of human, left cheek, chin and forehead, it can input Need (6) on the left corner and finally push (7) Generate Result, which is indicted by Figure 3.4, which indicates (8) red collection point of input value, skin color ellipse will display this (9) ellipse and input (10) L illumination to observe its changes, and stores some skin color RGB and Office Excel file of Lab.

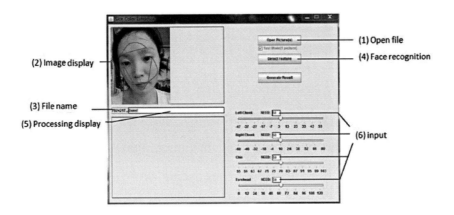

Figure 3.3   Instruction for SCE operation

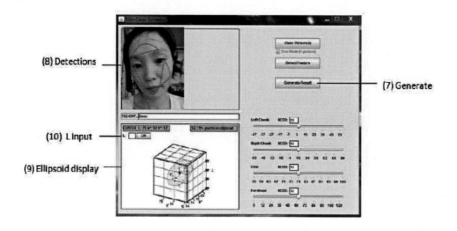

(8) Detections

(7) Generate

(10) L input

(9) Ellipsoid display

Figure 3.4   Processing by single image

3.1.2   Taguchi method finds optimization

Figure 3.5(a) indicates detection and positioning of human face is completed and connection line way by point to point. It makes division for right eye corner, left eye corner, middle of right eyebrow, left eyebrow, left mouth corner, right mouth corner. Figure 3.5(b) may be the area of producing skin color point, so Figure 3.5(c) indicates the probable color point, this research sets every inserted photo is 300x600 pixel, of which the pixel value of connection line, the maximum point of pull-up radian is 50 point, equivalent connection line of point to point is 25 point, the minimum pull-up radian is 3 point. Radian direction is respectively represented by -1, 0 and 1. For example, the minimum pull-up position of right mouth corner and left mouth corner will reach the chin of mouth and shadows, this radian direction cannot construct area block conforms to skin color, so it doesn't be listed into calculation.

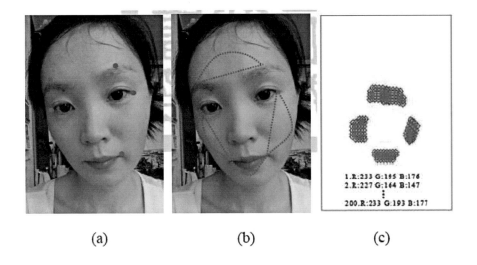

(a)　　　　　　　　(b)　　　　　　　　(c)

Figure 3.5　The distribution possibility for input in SCE.

Apply Taguchi method to get the optimization from the distribution possibility for input in SCE. Figure 3.6 shows the study divides face into 4 blocks; they are respectively forehead, left cheek, right cheek, and chin. It chooses proper factor as design level. One block respectively has factors (radian and point number), it totally has 8 factors, 3 grades, so it chooses L_{18} orthogonal table.

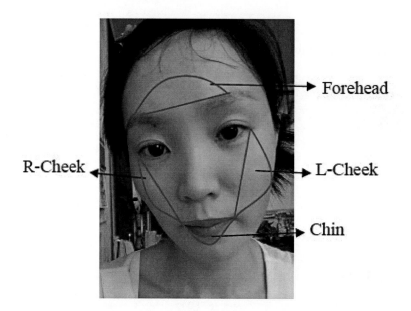

Figure 3.6   Definition and division of face part

Base on the 4 areas defined, Table 3.1 shows control factor table to clarify all parameters, and will follow the Taguchi method to do the test.

Table 3.1 A control factor table that may be generated for the colour detection.

Levels of Control Factors		Level	Level 1	Level 2	Level 3
Chin	radian	A	-1	0	
	points	B	3	25	50
R-cheek	radian	C	-1	0	+1
	points	D	3	25	50
L-cheek	radian	E	-1	0	+1
	points	F	3	25	50
Forehead	radian	G	-1	0	+1
	points	H	3	25	50

Using It uses characteristics of orthogonal table to reduce test times from 4,374 time to 18 times, it greatly simplifies test times and calculate S/N proportion, standard difference, average, it presents result by inputting into the table, through conversion of Taguchi method S/N proportion, which makes test data conforms additive model(addition characteristic), and calculates level effect of every factor, and gets factor reaction table of quality characteristic, which is indicate by Table 3.2.

Table 3.2   Factor reaction table

Exp	A	B	C	D	E	F	G	H	P1	P2	P3	Ave	S	S/N
1	1	1	1	1	1	1	1	1	98	97.5	98	97.83	0.289	50.6
2	1	1	2	2	2	2	2	2	94	93.8	94.2	94	0.2	53.4
3	1	1	3	3	3	3	3	3	95	94.5	96.2	95.23	0.874	40.7
4	1	2	1	1	2	2	3	3	96.7	96	95	95.9	0.854	41
5	1	2	2	2	3	3	1	1	94	95	94	94.33	0.577	44.3
6	1	2	3	3	1	1	2	2	94.4	94	94.5	94.3	0.265	51
7	1	3	1	2	1	3	2	3	96.2	96	96.5	96.23	0.252	51.7
8	1	3	2	3	2	1	3	1	96	96.5	96	96.17	0.289	50.5
9	1	3	3	1	3	2	1	2	96	96	95	95.67	0.577	44.4
10	2	1	1	3	3	2	2	1	95.4	95.4	96.7	95.83	0.751	42.1
11	2	1	2	1	1	3	3	2	95	94.5	95.2	94.9	0.361	48.4
12	2	1	3	2	2	1	1	3	96	95.1	96	95.7	0.52	45.3
13	2	2	1	2	3	1	3	2	95.4	94	96.1	95.17	1.069	39
14	2	2	2	3	1	2	1	3	96	94	95	95	1	39.6
15	2	2	3	1	2	3	2	1	95.1	96.7	96	95.93	0.802	41.6
16	2	3	1	3	2	3	1	2	95	94.5	95.8	95.1	0.656	43.2
17	2	3	2	1	3	1	2	3	94.5	96.6	96.5	95.87	1.185	38.2
18	2	3	3	2	1	2	3	1	96.7	94.5	96	95.73	1.124	38.6

From action table it can clearly see the effect result of quality characteristic, because S/N proportion belongs to projection characteristics, it can easily find the optimized result of every group in the table. Firstly, part of Significant in Table 3.3 can see B D F H, it is yes, and representative has reaction and effect. They are respectively: optimal combination of collection point is {BFHD}. It can also find factor significance sequence under quality characteristic, the optimized efficiency is B>F>H>D, such as table 4-5 factor characteristic result figure. Fraction of Rank, the first one is B, range from -4~1.7, range value of carry number is about 6, so optimal collection point is 6 point.

From analysis result, it finds that importance sequence will change according to quality characteristic, it mainly because Taguchi method belongs to the optimal method of single quality characteristic and then uses this to program correction base, which can make program of this research quickly calculate the optimized result of skin color collection.

Table 3.3   Factor characteristics result table

	A	*B*	*C*	*D*	*E*	*F*	*G*	*H*
Level 1	44.0	46.8	44.6	44.0	46.6	46.0	44.6	46.0
Level 2	41.8	42.7	45.7	42.0	45.8	43.2	44.0	44.0
Level 3		44.4	43.6	41.0	45.0	42.0	43.0	42.7
E^{1-2}	-2.2	-4.0	1.1	-2.0	-0.8	-2.8	-0.6	-2.0
E^{2-3}		1.7	-2.1	-1.0	-0.8	-1.2	-1.0	-1.3
Range	2.2	5.7	2.2	3.0	1.6	4.0	1.6	3.3
Rank	5	1	6	4	7	2	8	3
Significant	no	yes	no	yes	no	yes	no	yes

From analysis result, it finds that importance sequence will change according to quality characteristic, it mainly because Taguchi method belongs to the optimal method of single quality characteristic and then uses this to program correction base, which can make program of this research quickly calculate the optimized result of skin color collection.

3.2　Verification for 6 points to detect facial colour

FaceRGB mainly makes human face identification by digital image and makes point color task by characteristics collection and relative position. In order to demonstrate program reliability and practice, this research uses human spectrometer as skin color segmentation to make comparison.

3.2.1　Spectrometer Application

The scientific instrument of decomposing light into spectral line, it is composed of prism or grating diffraction, when polychromatic light makes light split by light split components(such as grating, prism), the graph formed by sequence of wavelength of light(or frequency), while the biggest part in the light spectrum, while the visible spectrum in light spectrum is only one part that can be seen by human eyes, the electromagnetic radiation in this wavelength range is called as visible light (James & John, 2007). This research uses USB4000 as experiment and research instrument, appliance for test demonstration, instrument respectively connects with receiver, sampling element, detector and fiber line. Finally it is connected to computer by USB connection line. (Ocean Optics Int Web site)

3.2.2 Test Design

1. Test object: Asian female aged from 15 to 35, sample number: 40 Figure 3.7 shows 40 sample for the test.

2. In the closed space of 2M cubic meters, Figure 3.8 is the space and related allocation design of research test, its wall uses matt black paint coating and back curtain to prevent light interference.

3. Design of carriage, on one hand it is used to fasten chin, on the other hand it needs to guarantee light (D65) and receiving angle of reflected light.

4. Spectrometer, light and other adjustable appliance and equipment all comes from Ocean Optics, its reliability and academic certification is acknowledged by the industry.

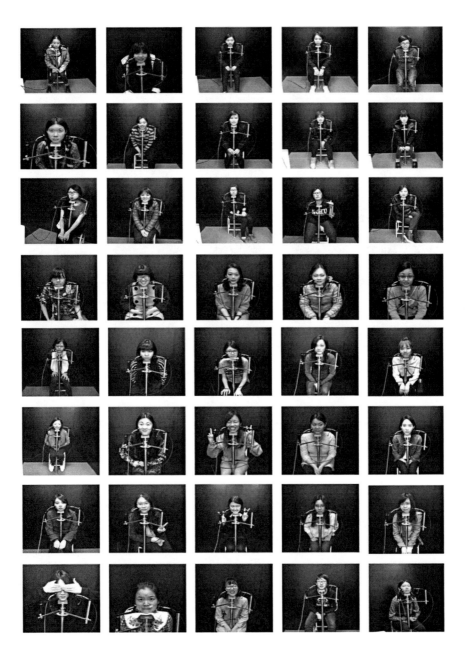

Figure 3.7　40 samples in the FOS test

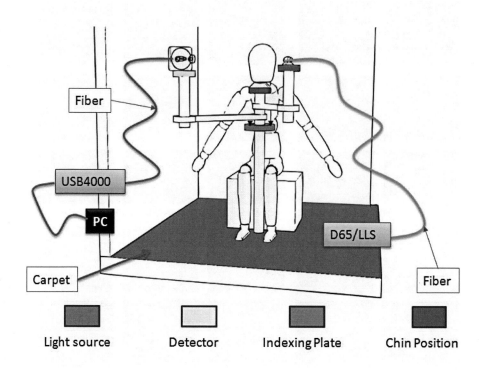

Figure 3.8   Design and layout for FOS test

3.2.3   Test comparison of FaceRGB and FOS (Fiber of spectrometer)

Through test comparison, Figure 3.9(a) is the digital image of FaceRGB using 40 tested samples and it uses color data by using 6 point color, research uses data as point data distribution process of space image, Figure 3.9(b) uses spectrometer focuses on 40 samples and gets data distribution in designed space by making actual measurement.

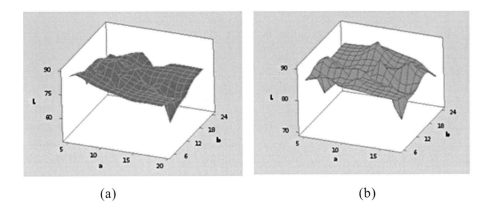

(a)　　　　　　　　　　　　　(b)

Figure 3.9　Color result of FaceRGB and spectrometer (FOS)

Reliability is the measurement task in measuring research, whether it can maintain uniform index and parallel forms reliability is the same test used in test and different forms, its similarity or equivalence property, it usually uses Pearson correlation coefficient to measure parallel reliability, referring to Eq.(16) the formula is as follows:

$$r = \frac{\sum(x-\bar{x})(y-\bar{y})}{\sqrt{\sum(x-\bar{x})^2}\sqrt{\sum(y-\bar{y})^2}} = \frac{n\sum xy - \sum x - \sum y}{\sqrt{n\sum x^2 - (\sum x)^2}\sqrt{n\sum y^2 - (\sum y)^2}}$$

(16)

Fiber of Spectrometer, FOS. Table 3.4 is the correlation coefficient of FaceRGB-L and FOS-L, its r= 0.962, P value =

0.763. Correlation coefficient of FaceRGB-a and FOS-a, r= 0.865, P value = 0.499, correlation coefficient of FaceRGB-b and FOS-b Pearson, r = 0.992, P value = 0.621. They all belong to strong correlation.

Figure 3.10 shows interval correlation of LAB, standard difference value of lateral axis, vertical axis: presentation of correlation value in correlated interval. Condition and difference of sample distribution is not obvious, which represents every sample point has close relations and interval has overlap, and it is placed on the same line, correlation strength is 95%.

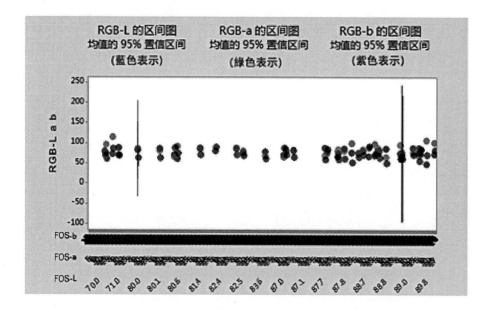

Figure 3.10   Interval correlation of Lab between FaceRGB and FOS test

Table 3.4　Using Minitab to make correlation calculation

相关: RGB-L, FOS-L

RGB-L 和 FOS-L 的 Pearson 相关系数 = 0.962
P 值 = 0.763

相关: RGB-a, FOS-a

RGB-a 和 FOS-a 的 Pearson 相关系数 = 0.865
P 值 = 0.499

相关: RGB-b, FOS-b

RGB-b 和 FOS-b 的 Pearson 相关系数 = 0.992
P 值 = 0.621

📊 工作表 2 ***

↓	C1	C2	C3	C4	C5	C6
	RGB-L	RGB-a	RGB-b	FOS-L	FOS-a	FOS-b
17	83.0000	12.4000	10.0000	83.6	12	10
18	85.0000	6.0000	10.0000	87.1	5	10
19	87.0000	6.2300	11.0000	88.8	6	11
20	82.0000	10.0000	14.5000	80.1	10	15
21	80.2500	10.0000	11.0000	81.4	10	10
22	89.1400	8.0000	6.0000	89.0	8	6
23	81.5970	6.5000	16.5000	82.4	7	16
24	88.2100	6.0000	12.8400	89.8	6	12
25	82.5000	6.5000	17.5000	82.5	7	17
26	89.8000	7.0000	13.5000	89.8	6	13
27	87.6200	6.8000	8.0000	87.8	7	8
28	88.1000	10.2000	9.0000	87.7	11	9
29	73.0000	7.3000	10.0000	71.0	7	9
30	86.8800	12.0000	16.0000	88.7	12	16
31	88.2300	10.5000	11.0000	89.8	10	10
32	87.0000	7.0000	5.0000	87.8	7	6
33	79.2000	9.2000	15.0000	80.0	9	15
34	87.2600	7.0000	10.8000	88.7	7	10
35	86.2400	7.0000	7.0000	88.8	7	7
36	87.1223	12.7334	11.5400	87.0	12	10
37	89.4443	8.0767	11.0000	89.0	8	11
38	81.4790	7.0777	16.2000	80.6	7	16
39	71.9288	14.9151	14.5600	70.0	15	14
40	90.8632	8.9037	11.0000	89.0	8	11

CHAPTER 4
CASE STUDY

After applying the Taguchi method to do the optimizations based on SCE program, the study make the verification for 6-points skin color detection hypothesis. It also has been verified with FOS test.　And author used the concept to create program FaceRGB which was used JAVA language as a platform, it could automatically process the entire dataset in the study.

4.1　FaceRGB

Six points have been selected based on the previous method, and there are three steps in the program:

1. Firstly, perform face recognition and extract facial-feature locations. Figure 4.1 (a) shows the feature points after face recognition.
2. The relative locations of the facial features are used to determine the six points.

3. Identify the colour from the six points.

This procedure selects identification points after feature recognition, through calculation, and uses the relative positions of these points to determine the six locations. Figure 4.1(b) shows the feature points, described by their relative positions: six distributed on the forehead (6), four points around the upper and lower cheek (1-4), and one point on the chin(5). The six captured points will be different in different images since the characteristics of the human faces are different.

(a)   (b)

Figure 4.1   From the face feature detection to colour detection distribution.

4.1.1　Outliers

In debugging, the purpose is to erase and filter out those points that do not represent the face colour and are very different from the six captured points. This paper uses average facial colour as a reference, and calculates the distance between each face pixel and the reference point. If the distance is too large, the point will be considered an outlier. In the preparation process of the program, the points that represent facial skin colour are captured, and then the mean value is calculated. The average distance is calculated and the standard differential displacement of the Gaussian distribution is used to then delete the outliers.

In this study, the proposed colour range (Hsu et al., 2002) is based on a statistical result from the literature. For the YC_bC_r colour space, pixels with C_b values from 97.5 - 142.5 and C_r values from 134 ~ 176 can be considered skin-color pixels; the rest are non-skin pixels.

Any of the six points may come from part of the hair or a shadow, since they are in the identification range of facial colour values but differ in brightness. For efficient debugging, besides the limited value, the Gaussian distribution concept and the standard deviation are used to determine the outliers. Figure 4.2

describes how to define and find the outliers from the six points.

The procedure of the FaceRGB program is as follows.

1. Calculate the facial skin data. For each point, calculate the average distance to $FaceLAB_{avg}$ (ΔE_{avg}) and the standard deviation σ.

2. Determine the outliers, for which the distance from $FaceLAB_{avg}$ is large (i.e., greater than $Distance_{avg} + 2\sigma$).

3. According to CIE2000 (Sharma et al., 2004), the formula given in Eq.(17) is applied.

$$\Delta E_{00}^* = \sqrt{\left(\frac{\Delta \acute{L}}{k_L S_L}\right)^2 + \left(\frac{\Delta \acute{C}}{k_C S_C}\right)^2 + \left(\frac{\Delta \acute{H}}{k_H S_H}\right)^2 + R_T \frac{\Delta \acute{C}}{k_C S_C} \frac{\Delta \acute{H}}{k_H S_H}}$$

(17)

4. Delete the outliers from the six points.

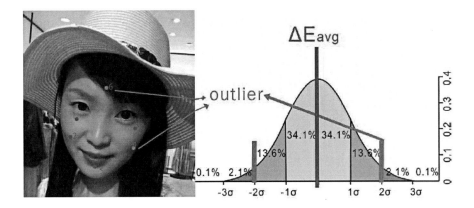

Figure 4.2 Apply the Gaussian distribution to the outlier concept.

4.1.2 FaceRGB operation

In this study, the FaceRGB program, based on the theories and methods described above, is integrated into the process automation. Figure 4.3 shows the interface of the program, which is divided into 6 display areas. They are described individually as follows:

1. When the program is opened, the title of the program, FaceRGB, is displayed.

2. When the file has been read, the image will appear in the picture window, and the status of big-data reading or other operation will be indicated. The instant synchronization status is shown in the window.

3. Spread sheets describing the progress are displayed in strip windows. The situation will progress according to the long schedule, and the results are presented as they are obtained.

4. For big data, the authors created four computation channels in the program to simultaneously deal with the huge data. Figure 4.4 shows the statuses of the channels as they are working.

5. The option is given one time to read the data as a single image or input.

6. A single image-processing result is displayed, including the colour, RGB values and LAB values. Figure 4.5 presents the example.

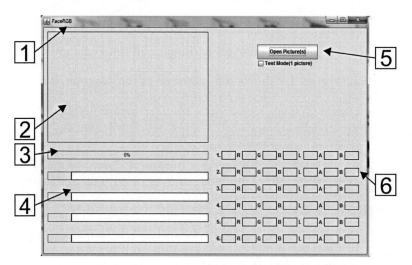

Figure 4.3   The interface of FaceRGB

Figure 4.4   The four channels for big data computation in FaceRGB

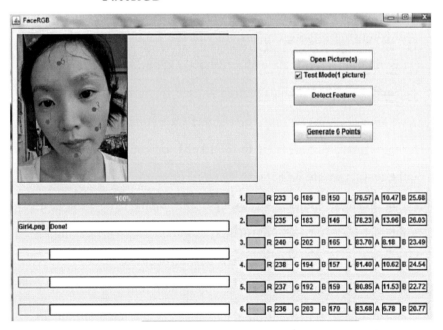

Figure 4.5   The processing for single image in FaceRGB

4.2 ColorFCM

To implement the next step, over 10,000 female images were collected, and the FaceRGB program was used to capture human skin colour. The size of the available data is 10,096 and more than 10,000 LAB colour data are generated. The concept of Fuzzy theory is integrated into the ColorFCM program for clustering.

4.2.1 Expectation Maximization

Expectation Maximization (EM) (Dempster et al., 1997) algorithms were used to determine the number of clusters (K).

1. EM considers three dimensions via the LAB-data probability distribution to calculate the overall likelihood function in K groups. The likelihood function is used for some known results obtained by observation, to estimate the parameters. For group representatives, the value of the likelihood could be higher and the results would be better.

2. For a small number of groups, the number of groups is increased (k = 1 ~ N) until the likelihood stops increasing; at this point, the number of groups will be decided.

3. In mathematical statistics, the likelihood function is used to calculate the statistical model parameters for determining the likelihood13.

For the filtering and smoothing EM algorithms, a two-step procedure (Expectation and Maximization steps) is repeated.

4.2.2 E-step (Expectation step)

The Kalman filter or a minimum-variance smoother designed with the current parameter estimates is applied to obtain updated state estimates. In Eq.(18), the parameter x denotes the probability distribution for Z utilization conditions.

$$Q\left(\theta \middle| \theta^{(t)}\right) = E_{Z|\chi,\theta^{(t)}}[\log L(\theta; \chi, Z)] \tag{18}$$

When three pieces of information (a statistical model which generates a set χ of observed data, a set of unobserved latent data or missing values Z, and a vector of unknown parameters θ.) are given, the expected value based on the conditional distribution of Z given χ under the current estimate of the parameters $\theta^{(t)}$ is calculated.

4.2.3 M-step (Maximization)

Use filtered or smoothed state estimates with maximum-likelihood calculations to obtain updated parameter estimates. Refer to Eqs.(19a)-(19b), which use maximization to optimize the parameters.

$$\theta^{(t+1)} = \underset{\theta}{argmax} \, Q\big(\theta|\theta^{(t)}\big),$$

$$(19a)$$

$$F(q,\theta) = E_q[\log L(\theta;\chi,Z)] + H(q) =$$

$$-D_{KL}\big(q| \big|p_{Z|X}(\cdot\,|x;\theta)\big) + \log L(\theta;x), \qquad (19b)$$

Where q is an arbitrary probability distribution over the unobserved data z, $p_{Z|X}(\cdot\,|x;\theta)$ is the conditional distribution of the unobserved data given the observed data x, H is the entropy and D_{KL} is the Kullback–Leibler divergence (Hastie et al., 2001).

4.2.4 K-means++

Arthur & Vassilvitskii (2007) provides the approach used in this step to spread out the k initial cluster centers. The first cluster center is chosen uniformly at random from the data points that are being clustered. After that, each subsequent cluster center is chosen from the remaining data points with probability proportional to its squared distance from the point's closest existing cluster center.

The exact algorithm is as follows:

1. Randomly choose one center from the uniformly distributed data points.
2. Compute D(x) for each data point x, which is the distance between x and the nearest chosen center.
3. The next step is to choose one new data point randomly as a new center and use a weighted probability distribution, where a point x is chosen with probability proportional to D(x)^2.
4. Repeat Steps 2 and 3 until k centers have been chosen.
5. Once the initial centers have been chosen, proceed with standard Fuzzy C-means (Arthur & Vassilvitskii, 2007).

Eq. (20) is the formula for the distance percentage. It is the key issue in the development of the program and will be the final decision.

$$P_D = \frac{D(x)^2}{\sum_{x \in X} D(x)^2} \tag{20}$$

where P_D is the percentage of the distance D(x)^2 relative to the total distance for all points.

4.2.5 ColorFCM implementation

The ColorFCM program for data clustering is based on three theories, EM, FCM and K-mean[++], and was created in the JAVA language. Figure 4.6 shows the ColorFCM program's interface. The following procedures are used for colour grouping:

1. The header: Items (Open the program, the header and ColorFCM) are listed on the upper-left corner of the screen.

2. Input: This procedure loads the FaceRGB program's results. There is 10,096 data points for female facial skin-color values. There are two options for grouping: one is manual, and the other is to specify the number of clusters for the program to calculate. It integrates until the required groups are found, the objective of FCM.

3. Do Clustering.

4. Get Result: A spreadsheet will appear to present the results; on the other hand, the classification of the LAB values also converts the results to the RGB model. Figure 4.7 shows the result of a clustering example.

5. 3D scatter plot: Designs 3D coordinates to display the distribution of three-dimensional colour-space data.

Figure 4.8 shows the 18 groups of the clustering result in 3D space.

6. Progress: It illustrates the progress of the strip calculations, the progress and results of the process status can be observed.

7. Main window: It depicts the process and results. The calculation status and messages can be seen.

8. After clustering is complete, the result can be applied to make-up. There are two ways to input facial images: from existing files in a device (PC, Smartphone or camera), and by taking a photograph with the webcam at this time.

9. Clear All Add: This is a specific function that is used if the wrong file has been input and allows operators to do some modification.

10. Get image(s) L*a*b*: This function captures the colour from the image input and converts the RGB values to L*a*b* values through the program.

11. Classification report: This allows the user to view the result of the last procedure and classify images into the big-data clusters.

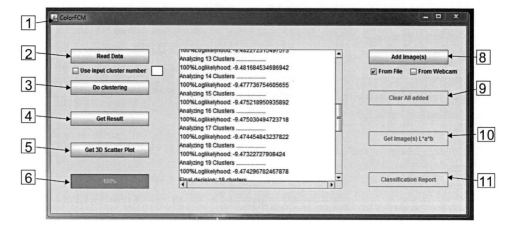

Figure 4.6 The interface of ColorFCM

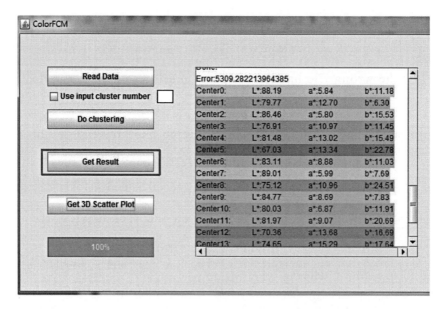

Figure 4.7 The result of ColorFCM with 10,096 data.

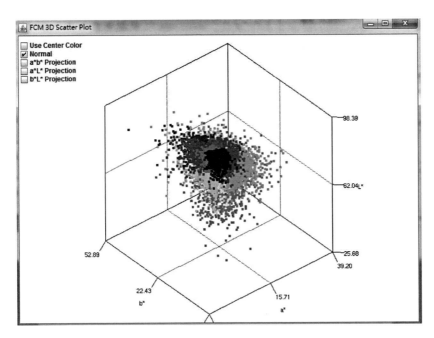

Figure 4.8　The 18 groups of the clustering result in 3D space.

CHAPTER 5
RESULTS & DISCUSSION

In this study, facial skin colour was captured via FaceRGB from six points; approximately 10,096 female images were used as data. The core objective of ColorFCM is to design a debugging mechanism. It assists in the removal of points representing non-face colours and organizes facial skin-color information. After FaceRGB, 10,056 data points were obtained and then grouped using fuzzy clustering theory. After that, for the ColorFCM program is executed, which combines the FaceRGB information obtained by the calculation program. Finally, 18 groups were obtained as the result. Figure 5.1 shows the entire process and the result.

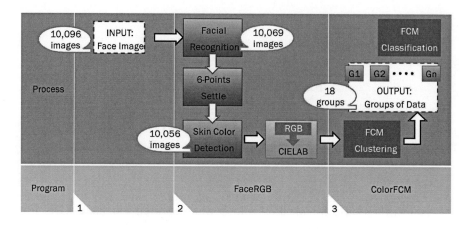

Figure 5.1   The entire process and result of the research in the study.

The results of the study can also be used for facial reference in this program with logical operators. Skin-color research from other countries or regions could also be adopted, to take into consideration local cosmetic preferences and skin colours. The design-automation program is used for computing and organizing big data, but can also be applied to daily photographic records of analysis and observation. Therefore, the huge amount of data that comes from daily cumulative records could be analyzed, and the data calculation could be repeated to generate new processes. In this way, such information is likely to increase the reference value.

5.1　Experimental verification

As in many studies related to skin colour, light and environmental factors influence the result of skin-color illustration in images. To verify the reliability of the FaceRGB program, female face in different surroundings is used for the experiment. Images of a woman in different surroundings (indoors, outdoors and dark background) are considered. Table 5.1 shows a digital camera taking pictures of a face in different lighting environments; six modes were captured using the FaceRGB program.

Table 5.1 The result of FaceRGB with one face

	Image		Points	R	G	B	R.avg	G.avg	B.avg	L*	A*	B*
C1			1	236	208	194	220	185	159	77.5	8.7	17.9
			2	219	184	156						
			3	217	182	154						
			4	217	173	138						
			5	203	163	127						
			6	231	204	187						
C2			1	219	175	140	219	180	149	76.0	9.7	21.1
			2	228	192	166						
			3	208	158	121						
			4	223	189	161						
			5	208	165	123						
			6	230	203	184						
C3			1	223	173	140	222	181	151	76.6	10.6	20.9
			2	230	194	172						
			3	211	162	119						
			4	221	186	158						
			5	218	169	137						
			6	231	204	183						
C4			1	171	127	80	169	128	91	56.7	11.0	26.2
			2	140	97	62						
			3	230	207	189						
			4	157	104	70						
			5	130	81	41						
			6	188	155	104						
C5			1	177	133	88	166	127	91	56.1	10.3	25.4
			2	137	98	59						
			3	215	198	182						
			4	150	101	61						
			5	137	89	51						
			6	182	144	105						
C6			1	165	125	76	154	113	73	50.9	11.3	28.4
			2	116	80	48						
			3	201	167	130						
			4	169	121	73						
			5	121	77	40						
			6	154	112	74						

5.2 RGB & YC_bC_r conversions

During the process of FaceRGB, the RGB model is firstly transformed to the CIE L*a*b* colour space. While "L *"denotes the brightness (luminance), with the range from 0 to 100, the value "a *"describes the green colour, with the range from -500 to 500, and "b *" describes the blue-to-yellow colour, with a range from -200 to 200. The centre of the circle represents lower saturation (Sharma et al., 2004). Usually images are saved in RGB colour-space format; however, this format is sensitive to light. To avoid the influence of brightness on the result, the image data are converted to the YC_bC_r colour space, which is less susceptible to being affected by luminance, to remove the skin from the image. YC_bC_r is a colour space that encodes a non-linear RGB signal; Y represents the luminance component and C represents the colour. Video images are usually used for continuous processing or digital imaging systems. The luminance Y represents the concentration of light and is nonlinear; a gamma-correction coding process is used. The C_b and C_r concentrations are the blue and red offset components.

$$\begin{bmatrix} Y \\ Cb \\ Cr \end{bmatrix} = \begin{bmatrix} 16 \\ 128 \\ 128 \end{bmatrix} + \frac{1}{256} \begin{bmatrix} 65.481 & 128.53 & 24.966 \\ -37.79771 & -74.203 & 112 \\ 112 & -93.789 & -18.214 \end{bmatrix} \begin{bmatrix} R \\ G \\ B \end{bmatrix}$$

$$(21)$$

To verify the experimental data to test resistance, RGB to YC_bC_r colour-space conversion was performed using Eq. (21) (Sharma et al., 2004) and the results are given in Table5.2. It also lists the skin-color ranges for band C_r. Figure 5.2(a) indicates that in the YC_bC_r colour space, the skin-color range is 97.5 ~ 142.5 for C_b and 134 ~ 176 for C_r; C1-C6 in Figure 5.2(b) were confirmed in the experimental data obtained from the six regions of the face. The verification is performed in the experiment.

Table 5.2   RGB & YCbCr conversions

	Image	R.avg	G.avg	B.avg	L^*	a^*	b^*	Y	Cb	Cr
C1		220	185	159	77.5	8.7	17.9	192.5	104.1	143.5
C2		219	180	149	76.0	9.7	21.1	188.1	108.1	145.4
C3		222	181	151	76.6	10.6	20.9	189.8	106.8	148.2
C4		169	128	91	56.7	11.0	26.2	136.0	105.7	146.6
C5		166	127	91	56.1	10.3	25.4	134.6	110.0	145.7
C6		154	113	73	50.9	11.3	28.4	120.7	107.8	146.8

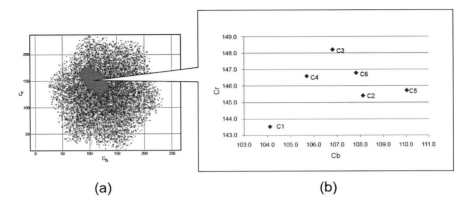

(a)　　　　　　　　　　　　　(b)

Figure 5.2　Skin Colour range

5.3　Training samples for FaceRGB

To verify that FaceRGB can be used for future analyses with big data, this study uses a pattern as a training database to perform the verification experiment.

1. Experimental objective: Use a smart phone (camera with a resolution of at least 600 megapixels) to randomly take 40 photos of women between the ages of 20 and 35. Colour data is used, and to enable prospective correction, the object is photographed under a variety of different environments and lighting conditions. Make sure that the women are not wearing makeup so that the skin-color data will be representative. Table 5.3 shows all the photos taken with the smart phone.

2. Experimental environment: Experiments were executed

under the Windows 7 operating system, and the FaceRGB program was written in the JAVA language; the hardware is a PC with Intel dual-core Core i3-350 M 2.4 GHz central processing unit (CPU), with a memory capacity of 8GB.

3. The results and analysis: Table 5.4 shows the details of 40 female face profiles with numbers, and shows each colour of the detection in different colour space. Three failed photos are recognized (02_NCKU, 21_NCKU, and 36_NCKU); in these photos, it is possible that the shooting angle is problematic, which means that the detection result would be influenced by facial features between the generated discernible results. When compared with Figure 4.2, Table 5.5 explains the details of the outliers. Verification of the program design was also performed.

Table 5.3　40 photo images in real time

Table 5.4　FaceRGB result of 40 images

Image	R.avg	G.avg	B.avg	L*	a*	b*	R1	G1	B1	R2	G2	B2	R3	G3	B3	R4	G4	B4	R5	G5	B5	R6	G6	B6
Sum	184	139	117	61.0	14.0	18.0	162	134	113	170	138	117	191	167	143	196	163	144	161	129	104	138	112	99
01_NCKU.jpg	169	140	120	60.4	7.9	14.8																		
02_NCKU.jpg	No Face Detected!																							
03_NCKU.jpg	199	138	129	63.3	22.1	14.2	248	184	174	184	120	111	226	168	157	178	118	110	168	103	101	190	137	121
04_NCKU.jpg	120	89	77	40.7	11.0	11.9	216	106	92	106	81	71	133	94	77	111	79	66	111	76	62	126	98	94
05_NCKU.jpg	200	150	123	66.2	15.2	21.5	200	166	141	205	153	131	197	157	122	183	136	108	204	147	118	195	143	119
06_NCKU.jpg	195	144	111	63.9	15.1	25.1	233	156	117	181	127	99	219	168	137	193	136	108	196	146	109	182	132	99
07_NCKU.jpg	215	162	136	70.9	16.2	21.1	155	200	181	206	148	110	230	180	155	210	146	88	200	137	130	N	N	N
08_NCKU.jpg	160	112	101	51.8	17.5	13.7	175	107	93	106	76	68	213	157	140	149	99	172	197	139	128	144	95	90
09_NCKU.jpg	206	177	165	74.4	8.6	10.0	191	146	130	182	146	134	244	220	208	218	181	104	214	182	169	208	188	179
10_NCKU.jpg	211	151	118	67.5	18.5	26.3	214	146	115	209	126	112	222	177	135	218	140	115	217	153	117	214	169	127
11_NCKU.jpg	195	144	125	64.2	16.7	17.7	223	163	146	179	126	110	205	157	135	184	131	123	191	136	116	201	153	133
12_NCKU.jpg	230	181	168	77.7	15.9	13.4	N	N	N	245	193	180	243	205	186	197	133	175	214	166	152	252	212	200
13_NCKU.jpg	226	172	154	74.8	17.3	16.9	201	169	143	168	125	106	252	196	169	241	187	87	145	147	137	240	191	177
14_NCKU.jpg	154	120	99	53.2	10.2	16.6	235	159	135	177	125	101	145	117	93	140	108	122	200	107	84	128	106	93
15_NCKU.jpg	211	161	128	70.1	14.2	24.3	165	175	138	152	139	121	222	173	140	200	149	149	214	166	128	223	179	144
16_NCKU.jpg	191	147	128	64.6	13.9	16.4	235	165	106	137	93	80	207	161	131	184	167	111	184	147	129	186	144	128
17_NCKU.jpg	195	149	122	65.4	13.6	20.9	202	188	158	164	118	103	218	168	107	140	137	85	140	104	124	195	158	132
18_NCKU.jpg	163	122	104	54.8	13.5	16.1	248	157	138	161	118	109	166	128	146	121	103	121	121	182	87	N	N	N
19_NCKU.jpg	191	150	134	65.4	13.0	14.3	186	210	210	183	146	127	201	166	139	197	163	135	172	144	113	189	146	129
20_NCKU.jpg	186	153	129	65.6	8.7	17.2		150	124				199	168							122	180	149	129
21_NCKU.jpg	No Face Detected!																							
22_NCKU.jpg	180	142	116	61.9	10.7	19.3	203	166	137	185	148	121	207	168	137	189	150	121	107	73	61	189	150	121
23_NCKU.jpg	206	134	108	62.7	24.7	25.4	217	143	116	177	110	84	252	187	159	189	113	90	201	130	98	200	125	102
24_NCKU.jpg	127	114	80	48.4	-0.3	20.7	143	135	99	129	115	78	N	N	N	137	122	81	127	103	79	100	95	66
25_NCKU.jpg	102	85	94	38.1	8.7	-2.6	103	85	85	107	107	98	N	N	N	105	88	98	N	N	N	80	84	95
26_NCKU.jpg	173	128	95	57.2	13.2	24.7	218	176	136	172	120	83	213	178	140	181	131	96	177	114	81	200	51	37
27_NCKU.jpg	197	130	103	60.6	22.8	25.2	234	165	132	167	106	87	217	144	103	156	94	79	208	137	107	185	137	104
28_NCKU.jpg	186	131	97	59.5	17.1	2.7	161	119	94	152	102	75	210	147	103	202	138	102	202	150	113	226	132	100
29_NCKU.jpg	220	185	189	78.2	13.2	15.5	226	197	202	219	178	184	226	190	194	224	183	189	155	158	157	194	207	211
30_NCKU.jpg	177	118	107	55.4	21.8	9.7	193	129	119	180	116	104	169	112	103	176	118	106	155	98	87	187	137	126
31_NCKU.jpg	191	165	153	69.7	7.6	12.2	200	177	194	224	166	155	222	203	196	183	155	141	179	131	121	184	161	146
32_NCKU.jpg	212	195	175	79.7	2.7	22.8	230	218	196	227	209	188	250	243	224	209	183	168	225	153	128	101	169	150
33_NCKU.jpg	214	166	135	71.8	13.5	27.7	253	205	169	164	162	130	254	231	190	228	161	132	165	160	130	166	78	64
34_NCKU.jpg	166	102	72	49.6	22.6	11.9	165	105	81	159	99	77	186	116	80	177	105	65	130	86	58	167	101	71
35_NCKU.jpg	155	132	116	56.8	6.2		170	154	129		130	114	152	134	114	152	125	114	130	110	99		141	126
36_NCKU.jpg	No Face Detected!																							
37_NCKU.jpg	206	133	87	62.2	23.4	36.2	254	204	141	173	94	63	244	169	104	163	89	50	191	108	76	212	134	88
38_NCKU.jpg	211	149	117	67.0	19.5	26.2	199	133	99	207	148	114	212	147	125	216	153	122	217	155	116	216	159	130
39_NCKU.jpg	126	92	62	41.9	10.0	22.7	129	94	66	131	90	60	139	108	79	140	106	68	94	62	39	N	N	N
40_NCKU.jpg	137	92	58	43.2	14.6	26.9	123	80	46	99	57	32	223	183	147	132	79	45	123	73	38	127	81	45

Table 5.5   The description of outlier

	Open Picture	Detect Feature	Generate 6 Points	Result
07_NCKU.jpg				1. R 233 G 200 B 181 L 82.90 A 8.75 B 13.81 2. R 206 G 148 B 110 L 66.06 A 17.31 B 28.70 3. R 230 G 180 B 155 L 77.18 A 14.70 B 19.70 4. R 216 G 146 B 108 L 66.01 A 19.82 B 29.79 5. R 200 G 137 B 130 L 63.16 A 23.19 B 13.51 6. R 122 G 73 B 43 L 36.13 A 17.80 B 26.50
18_NCKU.jpg				1. R 202 G 157 B 138 L 68.39 A 14.04 B 16.36 2. R 164 G 118 B 103 L 53.84 A 16.01 B 15.40 3. R 166 G 126 B 107 L 56.69 A 11.67 B 17.04 4. R 140 G 103 B 85 L 46.86 A 12.34 B 15.97 5. R 143 G 104 B 87 L 47.53 A 13.26 B 15.72 6. R 51 G 39 B 43 L 17.13 A 6.44 B -0.49
24_NCKU.jpg				1. R 143 G 135 B 99 L 56.16 A -2.89 B 20.48 2. R 129 G 115 B 78 L 48.86 A -0.30 B 22.47 3. R 160 G 151 B 118 L 62.44 A -2.12 B 18.62 4. R 137 G 122 B 81 L 51.65 A -0.45 B 24.56 5. R 127 G 103 B 79 L 45.33 A 5.95 B 17.16 6. R 100 G 95 B 66 L 40.06 A -3.01 B 17.35
25_NCKU.jpg				1. R 103 G 85 B 85 L 37.91 A 7.42 B 2.77 2. R 107 G 85 B 98 L 38.73 A 11.50 B -4.05 3. R 182 G 170 B 184 L 70.98 A 6.90 B -5.61 4. R 105 G 88 B 98 L 39.35 A 8.86 B -3.16 5. R 163 G 152 B 196 L 64.16 A 6.68 B -5.88 6. R 94 G 84 B 95 L 37.62 A 6.36 B -4.88

5.4   RGB with a large quantity of images

In this study, Authors focus on applying FaceRGB to images of Asian females for big-data analysis. The study designed as follows:

1. Experimental objective: Over 10,000 images of young women (20-35 years old) were collected from the network.

2. Experimental environment: Experiments were executed

under the Windows 7 operating system, and FaceRGB was created in JAVA. The hardware is a PC with Intel dual-core Core i3-350 M 2.4 GHz central processing unit (CPU), with a memory capacity of 8GB.

3. The results and analysis: Figure 5.1 shows the results on 10,096 collected images. After filtering, there were 10,069 images to be detected and 10,056 sets of data were obtained.

5.5　ColorFCM result by Fuzzy C-means

After collecting the 10,056 sets of data using FaceRGB as row data, another program, called ColorFCM, was used for clustering.

1. Experimental objective: Authors collected 10,056sets of data using FaceRGB; the data could be in the format of an Excel Table.

2. Experimental environment: Experiments were executed under the Windows 7 operating System, and ColorFCM was created in JAVA. The hardware was a PC with an Intel dual-core Core i3-350 M 2.4 GHz central processing unit (CPU), with a memory capacity of 8GB, and a webcam.

3. Data Clustering: Based on Eqs. (18), (19a), (19b), and (20),

the program "ColorFCM" was designed to group captured colours.

4. The results and analysis: Table 5.6 shows the clustering result from 10,056 images. There are 18 groups from the FCM processing. Table 5.6 presents the L*a*b* values and the colour bar to identify each group. Figure 5.3 shows the various types in a 3D scatter plot of ColorFCM. Figure 5.3(a) is the skin-color clustering distribution on the a*b* projection; Figure 5.3(c) is the skin-color clustering distribution on the L*a* projection; and Figure 5.3(e) is the skin-color clustering distribution on the L*b* projection. For visualization, because some skin colours are very similar, a different colour is used to identify each group and Figure 5.3(b)(d)(f) show the projection.

Different data sets can generate individual correlations for specific purposes, such as identifying business trends, preventing diseases, and combating crime. Scientists, business executives, and those working in media, advertising and governments regularly encounter difficulties with large data sets in areas including Internet search, finance and business informatics. Scientists encounter limitations in scientific work, in fields such as meteorology, genomics, connectomics, complex physics simulations, and biological and environmental research (Rechman et al., 2011).

Table 5.6   The groups of facial color in ColorFCM

Center	L*	a*	b*	Color
0	81.05	7.48	13.18	
1	88.47	5.34	9.97	
2	68.97	11.67	18.38	
3	81.04	10.78	14.54	
4	86.53	8.25	15.01	
5	72.6	15.36	16.28	
6	73.84	13.39	25.63	
7	87.14	5.42	8.69	
8	80.8	14.13	14.27	
9	76.75	10.87	14.28	
10	83.54	5.91	9.96	
11	83.46	8.84	11.45	
12	78.33	12.85	20.53	
13	73.79	14.64	10.93	
14	87.94	7.09	7.24	
15	72.53	14.76	23.23	
16	64.37	13.18	11.02	
17	83.54	9.18	8.37	

The study focuses on facial colour to explore the various types of bases from a big-data perspective. The main goal is to produce an important reference for cosmetic applications.

Users could have knowledge of their skin colours and the colour locations of the face region, which can assist them in selecting the appropriate colours to match their skin. With this method, females could more easily determine their skin-color groups. Furthermore, with colour harmony and applied aesthetics, every clustering result generated by the fuzzy clustering method could be a potential fashion trend in cosmetics. Finally, if this system could be applied in the make-up market, it would make a considerable contribution.

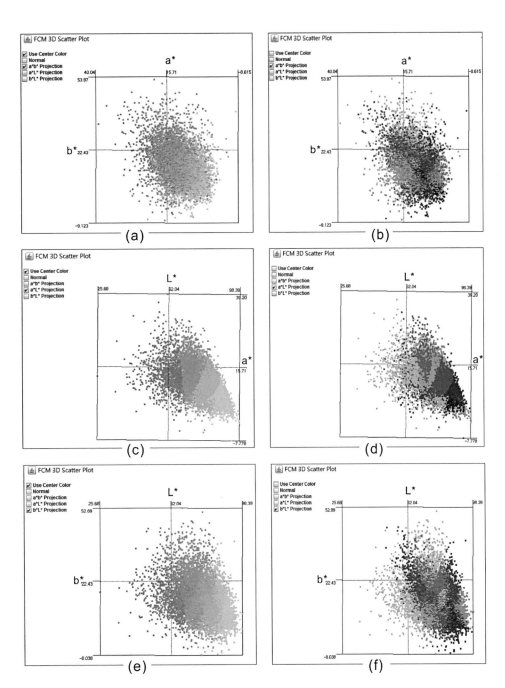

Figure 5.3　3D scatter plot of ColorFCM

CHAPTER 6
APPLICATION

There are six types of skin-color tone mentioned in Colour Forecast from Merck (2016). It also provides four trends: Metamorphoses, Design & Crafts, Neo-Realist and Freestyle (Rona, 2015). In this paper, the colour tone was detected and 6 clusters were assigned for processing with ColorFCM; Table 6.1 shows six colour tones: (1) very fair; (2) fair; (3) medium; (4) dark; (5) very dark; and (6) yellow undertones. The face-colour clusters that were previously obtained with ColorFCM can be connected with the trends. Table 6.2 shows the result. The authors also transform L*a*b* to RGB. Figure 6.3 presents the clustering distribution.

Table 6.1   Six color tones from Color Forecast (Rona, 2015)

Skin tones	RGB	LAB	CMYK	RGB	LAB	CMYK
		up			down	
Very Fair	R 254	L 96	C 1	R 223	L 81	C 16
	G 241	A 2	M 8	G 195	A 7	M 27
	B 218	B 13	Y 17	B 163	B 20	Y 37
			K 0			K 0
Fair	R 251	L 92	C 2	R 204	L 66	C 25
	G 227	A 6	M 15	G 145	A 21	M 51
	B 203	B 15	Y 22	B 122	B 22	Y 49
			K 0			K 0
Medium	R 249	L 85	C 3	R 194	L 62	C 30
	G 203	A 13	M 28	G 135	A 21	M 54
	B 169	B 23	Y 34	B 103	B 27	Y 59
			K 0			K 0
Dark	R 237	L 77	C 9	R 189	L 59	C 33
	G 178	A 18	M 38	G 128	A 21	M 57
	B 128	B 34	Y 51	B 88	B 32	Y 67
			K 0			K 0
Very Dark	R 228	L 65	C 13	R 156	L 42	C 44
	G 130	A 35	M 60	G 75	A 33	M 80
	B 72	B 48	Y 73	B 45	B 34	Y 92
			K 0			K 9
Yellow Undertones	R 246	L 86	C 6	R 190	L 62	C 32
	G 209	A 7	M 23	G 137	A 17	M 52
	B 148	B 35	Y 46	B 85	B 36	Y 70
			K 0			K 0

Table 6.2　Six color clustering from ColorFCM

Center	L*	a*	b*	Color	R	G	B
0	87.8	5.6	9.1		238.0	216.0	203.0
1	82.0	8.6	13.4		230.0	198.0	179.0
2	81.5	10.7	15.4		233.0	195.0	174.0
3	69.1	13.4	16.3		203.0	159.0	140.0
4	79.5	9.9	10.2		223.0	190.0	178.0
5	77.6	13.4	19.2		229.0	182.0	157.0

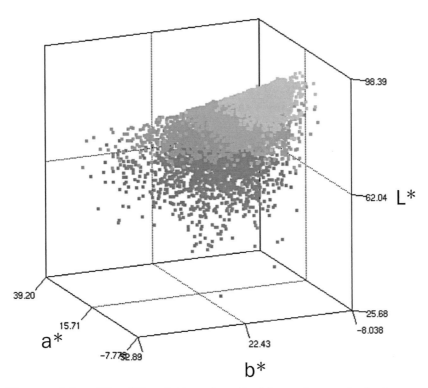

Figure 6.3　**The 3D plotter for 6 skin colour clustering distribution**

6.1 Connection between the colour clusters and Merck makeup trends

The relationship between the facial colours and the Merck makeup trends was established using the distance percentage (P_D) defined in Eq. (20).

Table 6.4, in which the clustering and fashion colours are included, shows the results of P_D for all colours. The result could be used to suggest colours that match the Merck makeup trends to customers.

Table 6.4   The Percentage of distance (P_D) between trend's & clustering's colour

		P_D
Very Fair		0.9637
Cluster 3		0.9904
Fair		0.9177
Cluster 6		0.9362
Medium		0.9592
Cluster 2		0.9848
Dark		0.7235
Cluster 1		0.9219
Very Dark		0.6439
Cluster 4		0.5339
Yellow		0.8329
Cluster 5		0.9270

6.2　For personal application

The user can input his/her photos into the system to capture face colours with ColorFCM.　The result of an example is shown in Table 6.5. In this table, user image, facial colour, Merck makeup trends, and the distance percentage (P_D) for the relationship between the customer's face colour and the Merck trends are all presented. The result shows that the face colour fits the medium colour tone. Based on this information, the user can select suggested cosmetic colours to match the face colour.

Table 6.5　Classification of the personal face color by the Percentage of distance (P_D)

Photo	P_D	L*	a*	b*	Color Chip	Corresponding color tone
	1.0	78.2154	8.427296	12.61771		Very fair
	1.0	83.14375	10.45294	12.84355		Fair
	1.0	71.10794	13.80814	20.80092		Medium
	0.5438	66.15316	15.20902	21.54749		
	1.0	72.49629	15.28242	13.4886		Dark
	1.0	75.84692	11.35917	18.76176		Very dark
	1.0	86.9966	6.234462	11.2712		Yellow undertone

6.3 For trend applications

In the manufacturing and marketing of cosmetics, makeup and skin-color matching is the focus of fashion-creation trends. Focusing on the cosmetics demand and the colour makeup trends, Merck (2016) proposes the makeup-trend forecast annually. According to the results of this study, face colours are classified into six clusters, which can be linked with Merck's trend forecasting. Table 6.6 shows the four trends forecasted by Merck (2016) for 2016-2017: Metamorphoses, Design & Crafts, Neo-Realist and Freestyle. They are two looks for each trend, and Rona (2015) also provides 6 types of face colours to apply with the trends. In this study, 6 skin colours are generated from ColorFCM in Table 6.2. Table 6.4 presents all skin colours applied with different trends. It shows six skin tones: dark, fair, very fair, dark, very dark, medium and yellow undertones. Table 6.8 shows the identifications for all looks. A user can determine her skin-color tone using a web camera, FaceRGB, and ColorFCM, and then connect her tone with the trends shown in Table 6.7 to preview makeup results and to see how makeup colours connect with the trends in Table 6.8.

The relationships among the skin-color tones, makeup, and

looks are shown in Table 6.7. With this table, the user can decide how to apply makeup based on her skin colour. The ways in which to apply makeup to fit the looks for all different skin-color tones, including very Fair, Fair, Medium, Dark, Very Dark, and yellow undertone, are described as follows, and the details are given in Table 6.8.

Table 6.6　Trend Forecast for 2016-2017 by Merck(2016)

Trends	Image	looks	Description
Metamorphoses		Look 1.	Animal Instinct
		Look 2.	Antique Goddess
Design & Crafts		Look 1.	Modeling
		Look 2.	Universal Glamour
Neo-Realist		Look 1.	Norm-Cute
		Look 2.	Magic Simplicity
Free Style		Look 1.	Party Girl
		Look 2.	Electric Impulse

Very Fair: The nude beauty mood is accentuated with precious shine cues in ANTIQUE GODDESS. The complexion is slightly whitened and cheeks are warmed up with coppery blush. Soft brown eye shadows create the light smoky effects seen in MODELING. Silver-white metallic accents infuse a fresh graphic feel and modern touch of light.

Fair: A fresh and graphic mood is inspired by NORM-CUTE with striking black or blue colour-blocked eyes. A white liner effect on the lower lips creates a new feel. Lips are nude or one tone darker. The complexion is clear and satiny, as seen in ANTIQUE GODDESS. Table 6.6 shows the images for the norm-cute and antique goddess looks.

Table 6.7   The face colour with trend

Skin Tones	Make-up										Looks
Very Fair	complexsion & cheeks		complexsion & nails		lips		eyes				Antiques Goddess Modeling
Fair	complexsion		eye & brows		lips		eye & nails		eyes		Antiques Goddess Norm-Cute
Medium	lips & nails		complexsion		eyes				eye & nails		Universal Glamour Animal Instinct
Dark	complexsion & cheeks		lips & nail		eyes				eye & nails		Party Girl Modeling Electric Impulse
Very Dark		eye & nails			lips				cheeks & eyes		Electric Impulse Modeling
Yellow Undertone	complexsion				lips & n				eyes		Magic Simplicity Modeling

Medium: This mood contrasts cold and warm tones. Eye shadows are matte, in blue and mauve, as seen in the palette of UNIVERSAL GLAMOUR, while intense dark red for lips pick up on ANIMAL INSTINCT. The subtly satiny complexion references UNIVERSAL GLAMOUR. Table 6.6 shows the images for the universal glamour and animal instinct looks.

Dark: This colourful and rich beauty mood is inspired by PARTY GIRL. Eyes are colour blocked in deep purple or midnight blue and highlighted with metallic reflections or sequins for evening. The complexion is warmed up by the ambery caramel blush of MODELING. Lips are wind red and subtly iridescent, as seen in ELECTRIC IMPULSE. Table 6.6 shows the images for the modeling and electric impulse looks.

Very Dark: A flamboyant and very feminine mood is inspired by ELECTRIC IMPULSE. Eyes and nails are decked in metallic emerald green, and lips in flashy fuchsia pink or intense purple. Cheeks are lightly shaded, as seen in MODELING.

Yellow Undertones: This fresh pastel look inspired by MAGIC SIMPLICITY features transparent sky blue eye shadow and soft pink lip gloss. Perky colour accents wake up the look: a touch of acid lime green on eyes or violet plum on lips. Textures are shiny and transparent and the complexion light and fresh, like the Cushion Creams of MODELING. Table 6.6 shows the image for the magic simplicity look.

6.4 Conclusion

An intelligent skin-color capturing method is proposed in this article. With this method, the skin colour of a person's face can be detected using the six-point averaging method proposed in this study.

The collected facial colours for different subjects can be divided into several clusters using the fuzzy C-means method. The colour clusters can then be connected with the Merck makeup trends using the so-called distance percentage (P_D). The results could assist consumers in selecting the appropriate cosmetic colours.

There was much discussion about the effects of light and camera on image acquisition. This study does not address the differences in lighting conditions and devices when finding the solution. In a future extension of this study, we may use video recording at the cosmetics counter or another fixed location. Using a specific machine with a specific light for daily video recording, this extension will improve the performance of FaceRGB in a big data setting to produce results that are more representative and applicable.

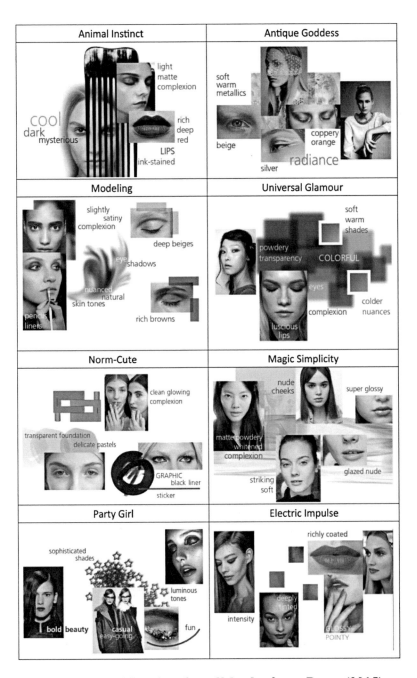

Table 6.8　Identification for all looks from Rona (2015)

REFERENCES

Arthur D, Vassilvitskii S. (2007). k-means++: The advantages of careful seeding. Proceedings of the 18th Anal ACM-SIAM symposium on discrete algorithms. SocIndAppl Math: 1027-1035.

Brunelli R, Poggio T. (1993). Face Recognition: Features versus templates. IEEE Trans on Pattern Analysis and Machine Intelligence;15:1042-1052.

Brunelli R. (2009). Template matching techniques in computer vision: Theo. Wiley.

Boughen N. (2003). Light wave 3d 7.5 Lighting. Word ware Publishing, Inc.

Bezdek JC. (2013). Pattern Recognition with fuzzy objective function algorithms: Springer science &business media. New York.

Caisey L, Grangeat F, Lemasson A, Talabot J, Voirin A. (2006). Skin color and makeup strategies of women from different ethnic groups. IntJ Cosmet Sci; 28:427–437.

Dempster AP, Laird NM, Rubin DB. (1997). Maximum likelihood from incomplete data via the EM Algorithm. J. R. Stat. Soc.; B39:1–38.

Duhn, S. von; Ko, M. J.; Yin, L.; Hung, T.; Wei, X. (2007). Three-View Surveillance Video Based Face Modeling for Recognition. Information Forensics and Security IEEE Transactions on Pattern Analysis and Machine Intelligence; 7: 1802-1811 .

Estanislao R, Suero M, Galzote C, Rivera Z, Li J, Kim HO, Thomas S, Khaiat A, Cheong EJ, Mangubat MI, Moideen R, TagamiH, Wang X. (2003). Characterization of Asian skin though in-vivo instrumental and visual evaluations: Influences of age, season and skin care habits. In: Proceedings of 6th Scientific Conference of the Asian Societies of Cosmetic Scientists, Manila; 263–269.

F'eraud R, Bernier OJ, Viallet JE, Collobert M. (2001). A fast and accurate face detection based on   neural

network. IEEE Trans on Pattern Analysis and
Machine Intelligence; 23:42-53.

Forsyth PH, Fleck MM. (1999). Automatic detection of
human nudes. Int J comput Vis; 32:63-77

Green P, Lindsay W, Donald M. (2002). Colour
Engineering: Achieving Device Independent
Colour. John Wiley and
Sons.　ISBN 0-471-48688-4.

Hsu RL, Mohamed AM, Jain AK. (2002). Face detection
in color image. IEEE Trans Pattern Anal; 24:696–
706.

Hsiao SW, Chiu FY, Hsu HS. (2008). A computer-assisted
color selection system based on aesthetic measure
for color harmony and fuzzy logic theory. Color
ResAppl;19:202–213.

Hunter RS. (1985). Photoelectric color-difference meter. J
Opt SocAm. ;48:985-995.

Hastie T, Tibshirani R, Friedman J. (2001). The EM
algorithm: The Elements of statistical learning.
New York: Springer.

Hsiao SW. (1994). Fuzzy set theory on car-color design. Color Res Appl;19:202-213.

Hsiao SW. (1995). A systematic method for color planning in product design. Color Res Appl;20:191-205.

Hsiao SW, Chang MS. (1997). A semantic recognition-based approach for car's concept design. Int J Vehicle Design;18:53-82

Hardin, R.H., & Sloane, N. J. A. (1993). A New Approach to the Construction of Optimal Designs, Journal of Statistical Planning and Inference; 37: 339-369

Ikeda N, Miyashita K, Hakim R, Tominaga S. (2014). Reflection measurement and visual evaluation of the luminosity of skin coated with powder foundation. Color Res Appl; 39: 45–57.

James & John. (2007), Spectrograph Design Fundamentals. Cambridge University Press. ISBN 0-521-86463-1

Jones, A. B., & Smith, C. D. (2005). Curriculum and instruction: A primer. New York: Academic Press.

Jones MJ, Rehg JM. (2002). Statistical color models with application to skin detection. Int J Comput Vis;46:81-96.

Kakumanu P, Makroginnis S, Bourbakis N. (2007). Asurvey of skin-color modeling and detection methods. Pattern Recognit; 40:1106-1122

Kumar C, Bindu A. (2006). An efficient skin illumination compensation model for efficient face detection IEEE IndEle IECON Anal:34444-34449.

Lam EY. (2005). Combining Gray World and Retinex Theory for Automatic White Balance in Digital Photography, IEEE Transactions on Consumer Electronics;14-16:134-139.

Mario D, Maltoni D. (2000). Real-time face location on gray-scalstaic image. Pattern Recognit;33:1352-1539.

Nock R, Nielsen F. (2006). On weighting clustering. IEEE Trans Pattern Anal;28:1–13.

Pukelsheim, Friedrich (2006). *Optimal Design of Experiments*. SIAM. ISBN 978-0-89871-604-7.

Reichman OJ, Jones MB, Schildhauer MP. (2011). Challenges and opportunities of open data in ecology, Sci.; 6018:703–705.

Rona, PECLERS PARIS. (2015). Merck Group.

Socolinsky, Diego A.; Selinger, Andrea (2004). "Thermal Face Recognition in an Operational Scenario". IEEE Computer Society; 1012–1019

Sharma G, Wu W, Dalai EN. (2004). The CIE DE2000 color-difference formula: Implementation notes, supplementary test data, and mathematical observations. ColorResAppl; 30: 21–30.

Smith, C. (2001). Understanding the role of instruction in the curricular context. In E. F. Johnson & G. H. Wise (Eds.); 210-230).

Smith, C. D., & Jones, A. B. (2004). Perspectives on curriculum and instruction. Journal of Educational Research; 55(1), 28-54.

Sun, Y., Wang, X., Tang, X. (2013). Deep Convolutional Network Cascade for Facial Point Detection. IEEE Conference on Computer Vision and Pattern Recognition ; 3476-3483.

Merck Group, Products & Industries. (2016). Available at http://www.merckgroup.com/en/products/products. html

Tan WR, Chan CS, Yogarajah P. (2014). Condell J, A Fusion approach for efficient human skin detection. IEEE Trans on IndInf;8:138-147.

Vadakkepat P, Lim P, Silva D, Jing L, Ling LL. (2008). Multimodal approach to human-face detection and tracking IEEE Trans Ind Ele;55:1385-1393

Wu H, Chen Q, Yachida M, (1999). Face detection from color images using a fuzzy pattern matching method. IEEE Trans on Pattern Analysis and Machine Intelligence; 21.

Yoshikawa H, Kikuchi K, Takata S, Yaguchi H. (2009). Development of avisual and quantitative evaluation method for facial skin color, Proceedings of 11th Congress of the AIC, CD-ROM, Sydney.

Yoshikawa H, Kikuchi K, Kim J, NishikawaS, YaguchiH, Mizokami Y. (2007). Effect of the chromatic components on the whiteness evaluation of skin.

Midtemmeeting of congress of international color
association; 203-206.

Zeng H & Luo MR. (2011). Skin color modeling of digital
photographic images. J. Imaging Sci. Technol;
55:030201

AI Application in Fashion Trend(英文版)
（人工智慧在流行趨勢研究的應用）

顏志晃(Chih-Huang Yen, Ph. D.) 著

發 行 人：賴洋助
出 版 者：元華文創股份有限公司
聯絡地址：100 臺北市中正區重慶南路二段 51 號 5 樓
公司地址：新竹縣竹北市台元一街 8 號 5 樓之 7
電　　話：(02) 2351-1607　　傳　　真：(02) 2351-1549
網　　址：www.eculture.com.tw
E-mail：service@eculture.com.tw
出版年月：2020 年 12 月 初版
定　　價：新臺幣 280 元

ISBN：978-957-711-195-1 (平裝)

總經銷：聯合發行股份有限公司
地 址：231 新北市新店區寶橋路 235 巷 6 弄 6 號 4F
電 話：(02)2917-8022　　　　傳 真：(02)2915-6275